# The Quantum Handshake

John G. Cramer

# The Quantum Handshake

## Entanglement, Nonlocality and Transactions

Springer

John G. Cramer
Department of Physics
University of Washington
Seattle, WA
USA

ISBN 978-3-319-24640-6 ISBN 978-3-319-24642-0 (eBook)
DOI 10.1007/978-3-319-24642-0

Library of Congress Control Number: 2015952772

Springer Cham Heidelberg New York Dordrecht London


Printed on acid-free paper

Springer International Publishing AG Switzerland is part of Springer Science+Business Media
(www.springer.com)

*This book is dedicated to Gilbert N. Lewis, John A. Wheeler, and Richard P. Feynman, who first envisioned the advanced-retarded handshake.*

# Foreword

Since its inception, quantum mechanics has been fraught with conceptual difficulties. What started as a noble attempt to understand the behavior of atoms, morphed into a set of working rules for calculating certain "observable" quantities.

These working rules have proven amazingly effective at dealing with a wide range of phenomena whose outcomes are statistical in nature, while attempts to understand the inner workings at a deeper level have met with immense frustration. My CalTech colleague, the late Richard Feynman, put it this way:

> One might still like to ask: How does it work? What is the machinery behind the law? No one has found any machinery behind the law. No one can explain any more than we have just explained. No one will give you any deeper representation of the situation. We have no ideas about a more basic mechanism from which these results can be deduced. [1]

All scientific advances require suspension of disbelief—if they followed directly from what came before, they would have already been discovered. The brilliant insights of the 1920's that led to our present situation were all of the form "If we use this particular mathematical representation, then the answer for the (…) comes out right", where (…) was, in many cases, certain spectral line energies of the hydrogen atom. While one must admire the fortitude of those who crafted these mathematical tricks, the result certainly does not constitute "understanding" in the sense that the term is used in the rest of science. Bohr and his followers attempted to remedy the situation with their "Copenhagen Interpretation", which has caused more confusion than the mathematics itself. As Ed Jaynes put it:

> …all these years it has seemed obvious to me as it did to Einstein and Schrödinger that the Copenhagen Interpretation is a mass of contradictions and irrationality and that, while theoretical physicists can, of course, continue to make progress in the mathematical details and computational techniques, there is no hope of any further progress in our basic understanding of nature until this conceptual mess is cleared up. [2]

Faced with this situation for what is now nearly a century, there have been two distinct positions taken by those in the field. The first is pragmatic, as articulated by Feynman:

**Fig. 0.1** Carver Mead (1934–), Gordon and Betty Moore Professor Emeritus of Engineering and Applied Science at the California Institute of Technology, was the originator of the term Moore's Law. He discovered that transistors would get faster, better, cooler and cheaper as they were miniaturized, thereby blazing the trail that has led to the microelectronics revolution

> Do not keep saying to yourself, if you can possibly avoid it, 'but how can it be like that?' because you will get 'down the drain', into a blind alley from which nobody has yet escaped. Nobody knows how it can be like that. [3]

Others believe that it will be understood, but may take a long time. John Archibald Wheeler said it well:

> Behind it all is surely an idea so simple, so beautiful, so compelling that when in a decade, a century, or a millennium we grasp it, we will all say to each other, how could it have been otherwise? How could we have been so stupid for so long? [4]

My own belief is that Wheeler is correct, but, instead of just one idea, a whole constellation of interlocking ideas must come together before the puzzle can be solved. High on the list of these ideas is the bidirectional arrow of time.

We humans live deep in the grips of thermodynamics, and all of our common experience is conditioned by it. To us it is natural that events in the past determine our situation in the here and now, but inconceivable that events in the future are affecting us in the present. So deeply is this conviction held that we never stop to ask from whence this asymmetry arises. The laws of electromagnetism are completely symmetrical with respect to both time and space. They always have two solutions:

1. A "retarded solution" that runs forward in time; and
2. An "advanced solution" that runs backward in time.

In spite of this symmetry, it is common practice, based on our thermal experience, to adopt the first solution and simply ignore the second. Already in 1909, Einstein had clearly stated the issue:

> In the first case the electric field is calculated from the totality of the processes producing it, and in the second case from the totality of processes absorbing it… Both kinds of representation can always be used, regardless of how distant the absorbing bodies are imagined

> to be. Thus one cannot conclude that that the [retarded solution] is any more special than the solution [containing equal parts advanced and retarded]. [5]

At the quantum level, things are quite different from our thermal world: When an atom in an excited state is looking for a way to lose its energy, it must find one or more partners willing and able to receive that energy, regardless of how distant they are. The Einstein solution, half advanced half retarded, creates a perfect "handshake" by which both atoms can accomplish their energy transfer.

In this book, John Cramer gives us a simple way of resolving many quantum mysteries by adopting the handshake as the common coinage of quantum interaction. It is by far the most economical way to visualize what is going on in these "mind-twisters" without departing from the highly successful mathematics of existing quantum mechanics.

Seattle
July 2015

Carver Mead

## References

1. R.P. Feynman, R.B. Leighton, M. Sands,*The Feynman Lectures*, vol. 3 (Addison-Wesley, Reading, 1965). ISBN 0201021188
2. E. Jaynes, Probability in Quantum Theory, in *Complexity, Entropy, and the Physics of Information* ed. by W. Zurek (Addison-Wesley, Reading, 1990), pp. 381–403
3. R.P. Feynman, The Character of Physical Law (MIT Press, Cambridge, 1967), p. 129
4. J.A. Wheeler, How Come the Quantum? Ann. New York Acad. Sci. **480**, 304–316 (1986)
5. A. Einstein, On the Present Status of the Radiation Problem (Zum gegenwärtigen Stand des Strahlungsproblems), Physcialische Zeitschrift 10 (1909); translated in A. Beck, P. Havas, (eds.), *The Collected Papers of Albert Einstein*, vol. 2, (Princeton University Press, Princeton, 1989)

# Preface

In 1900 there were clouds on the scientific horizon, unexplained phenomena that foretold the coming of the great intellectual revolution that was quantum mechanics. (See Chap. 2.) Today the scientific horizon is not without similar clouds. Quantum field theory, our standard model for understanding the fundamental interactions between particles and fields, tells us that the energy content of the quantum vacuum should be $10^{120}$ times larger than it actually is. At the interface between general relativity and quantum field theory, information seems to be vanishing at event horizons in a very unphysical way (see Sect. 6.21). Within black holes, singularities are predicted to exist that are completely beyond the reach of contemporary physics. Quantum chromodynamics, our standard model of particles, works very well in agreeing with high-energy physics experiments, but it employs two dozen arbitrary particle masses and interaction strengths. We have no idea where these values came from, how they are related, or how they were set in the early universe.

We can anticipate that our current understanding of Nature at the smallest and largest scales is at best a rickety scaffolding that must inevitably be replaced or improved. It is likely that another scientific revolution is on the way. Quantum mechanics will certainly play a key role in this revolution, but it is currently hampered by our lack of understanding of its inner mechanisms and our inability to visualize the many counter-intuitive aspects of quantum behavior. The Transactional Interpretation of quantum mechanics, presented and illustrated in this book, provides tools for visualization, for understanding quantum processes, and for designing new experiments. It paves the way for future theoretical and technical progress and in our understanding of the way the universe works. These tools should also be useful in the coming computation revolution, based on artificial intelligence, quantum computing, and quantum communication (see Chap. 8) that will, if properly used, lead to improvement of the human condition and benefit all of us.

This book gives an overview of the interpretational problems of quantum mechanics, provides an introduction to the Transactional Interpretation, and then demonstrates the use of it in understanding what is going on "behind the scenes" in many otherwise strange and mysterious problems of quantum optics. The target audience is the intelligent reader with some grasp of basic mathematics and a curiosity about quantum mechanics, what it is and how it works. We will not shy away from using occasional equations, but we will use them sparingly, and only when they are needed to make an important point.

The style of this book is intended to be wedge-shaped, starting easy and progressing to the somewhat more technical. It begins with a narrative style and introduces new concepts slowly and carefully. It builds up the basic conceptual framework of the Transactional Interpretation slowly, and then swings into action, applying it to a large collection of otherwise mysterious and counter-intuitive experiments that illustrate the curious behavior of quantum phenomena. This requires some mathematics, but the experiments that require the heavy use mathematics for their analysis are placed in Appendix D. The reader can "surf" over the mathematical parts and still gain a deep appreciation of what quantum mechanics is and how it works.

A certain mental flexibility will be asked of the reader in mastering the concept of "advanced" waves that are going backwards in time and carrying negative energy into the past. We are conditioned by the everyday world of experience to expect an "arrow of time" that always points from the past to the future, and we are disturbed by anything that seems to be going in the wrong time direction. However, the fundamental equations of physics have a time symmetry that recognizes no preferred time direction. A movie made of the behavior of fundamental particles looks OK, whether the images are presented in the time-normal or the time-reverse sequence. How this time-symmetric microcosm scales up to become the time-forward-only everyday world is a very deep question that is discussed in some detail in Chap. 9. The Transactional Interpretation, described in this book, uses waves going in both time directions with the two types doing handshakes as one of its basic quantum mechanisms, because that mechanism can be seen in the quantum formalism itself and because it allows us to understand the weirdness of entanglement and nonlocality. However, we note that these advanced time-running-backwards effects are limited to just the formation of time-forward transactions and are never allowed to produce "advanced effects" that would violate cause-and-effect.

The reader is also warned that there are a large and growing number of interpretations of quantum mechanics, of which the Transactional Interpretation is only one. I am reminded of a story that I heard long ago about a young child who was growing up in a house operated by the Berlitz School of Languages as a residence for the language teachers of the School. The child's mother spoke to him in English, his father in French, and the other occupants of the house each spoke to him in a different language. One day, the child began to speak in gibberish. His parents, after hours of persuasion, finally convinced him to talk to them in a language that they could understand. "Well," he said, "I'm getting to be a pretty big boy, and I decided that it was time that I had a language of my own." This is much the way it is with quantum interpretations: philosophers of science and philosophy-inclined physicists seem to prefer to have an interpretation of their own. Most of the resulting interpretational attempts have the problems that they introduce changes to the standard quantum formalism or they address only a restricted subset of the many interpretational problems and issues of the quantum formalism. In this book we will focus on the Transactional Interpretation and will only compare and contrast it with the orthodox Copenhagen Interpretation. We will discuss alternative interpretations only peripherally. It would require a much larger book (and one that I would not enjoy writing) to comprehensively present and criticize all of the many interpretations of quantum mechanics that are currently out there.

The portraits of physicists used in the opening chapters have been softened and rendered as "oil portraits" by using a conversion procedure on available photographs. The sources of the figures and portraits are specified in the **List of Figures** section on pages *xxiii–xxv*. There, the source notation "JGC" means that the figure was produced by the author.

Readers are encouraged to visit the Q&A section in **Appendix A** of this book whenever they encounter a term or idea that may require further clarification.

Seattle
July 2015

John G. Cramer

# Acknowledgements

The development of the ideas behind the Transactional Interpretation began in the 1980s and has continued for the past 35 years. During that period, the author has been greatly aided by discussions and correspondence with many people, some of whom are listed here alphabetically: John Blair, David Bodansky, David Boulware, Wil Braithwaite, Raymond Chiao, John Clauser, Kevin Coakley, Pauline Cramer, Paul Davies, Jorrit de Boer, Avshalom Elitzur, Dieter Gross, Ernest Henley, Nick Herbert, Max Jammer, Ruth Kastner, Emil Konopinski, Paul Kwait, Hans Krappe, Dan Larsen, Albert Lazzarini, Gerald Miller, Peter Mittelstädt, Rudolph Mössbauer, Chris Morris, Riley Newman, Rick Norman, Pierre Noyes, Kent Peacock, Rudi Peierls, Huw Price, Dennis Sciama, Saul-Paul Sirag, Henry Stapp, Derek Storm, Edward Teller, Dan Tieger, Lev Vaidmann, Carl von Weizsäcker, John Wheeler, Eugene Wigner, Larry Wilets, William Winkworth, and Anton Zeilinger.

I wish to thank the University of Washington's Center for Experimental Nuclear Physics and Astrophysics (CENPA, originally the UW Nuclear Physics Laboratory) for providing me with excellent research support and a stimulating intellectual atmosphere for over five decades. The author is also grateful to the following individuals, listed in alphabetical order, who have read and made valuable comments on preliminary versions of this book: Heidi Fearn, Munther Hindi, Carver Mead, Norbert Schmitz, Gudrun Scott, Ross Scott, William R. Warren, Jr., Ken Wharton, James Woodward, and especially to my wife, Pauline Cramer.

Seattle
July 2015

John G. Cramer

# Contents

**1 Introduction** .... 1
1.1 My Adventures with Advanced Waves and Time-Symmetry .... 1
1.2 Quantum Mechanics and Beer Bottles .... 6
References .... 8

**2 The Curious History of Quantum Mechanics** .... 9
2.1 Atomic Theory in the Early 20th Century (1900–1924) .... 9
2.2 Heisenberg and Matrix Mechanics (1925) .... 14
2.3 Schrödinger and Wave Mechanics (1926) .... 17
2.4 Heisenberg and Uncertainty (1927) .... 20
2.5 Heisenberg's Microscope (1927) .... 22
2.6 The Copenhagen Interpretation (1927) .... 24
2.7 Einstein, Podolsky, and Rosen; Schrödinger and Bohm (1935–1963) .... 28
2.8 Bell's Theorem and Experimental EPR Tests (1964–1998) .... 30
References .... 37

**3 Quantum Entanglement and Nonlocality** .... 39
3.1 Conservation Laws at the Quantum Scale .... 40
3.2 Hidden Variables and EPR Loopholes .... 42
3.3 Why Is Quantum Mechanics Nonlocal? .... 43
3.4 Nonlocality Questions Without Copenhagen Answers .... 44
References .... 44

**4 Reversing Time** .... 47
4.1 The History of *i* .... 47
4.2 Dirac and Time Symmetry .... 49
4.3 Wheeler–Feynman Absorber Theory .... 51
References .... 55

**5 The Transactional Interpretation** . . . . . 57
5.1 Interpretations and Paradoxes . . . . . 57
5.2 The One-Dimensional Transaction Model . . . . . 59
5.3 The Three-Dimensional Transaction Model . . . . . 62
5.4 The Mechanism of Transaction Formation . . . . . 66
5.5 Hierarchy and Transaction Selection . . . . . 67
5.6 The Transactional Interpretation of Quantum Mechanics . . . . . 68
5.7 Do Wave Functions Exist in Real 3D Space or Only in Hilbert Space? . . . . . 71
References . . . . . 73

**6 Quantum Paradoxes and Applications of the TI** . . . . . 75
6.1 Thomas Young's Two-Slit Experiment (1803)* . . . . . 75
6.2 Einstein's Bubble *Gedankenexperiment* (1927) . . . . . 78
6.3 Schrödinger's Cat (1935) . . . . . 80
6.4 Wigner's Friend (1962) . . . . . 82
6.5 Renninger's Negative-Result *Gedankenexperiment* (1953) . . . . . 83
6.6 Transmission of Photons Through Non-Commuting Polarizing Filters* . . . . . 85
6.7 Wheeler's Delayed Choice Experiment (1978)* . . . . . 88
6.8 The Freedman–Clauser Experiment and the EPR Paradox (1972)* . . . . . 91
6.9 The Hanbury Brown Twiss Effect (1956)* . . . . . 92
6.10 The Albert–Aharonov–D'Amato Predictions (1985) . . . . . 94
6.11 The Quantum Eraser (1995)* . . . . . 98
6.12 Interaction-Free Measurements (1993)* . . . . . 100
6.13 The Quantum Zeno Effect (1998)* . . . . . 105
6.14 Maudlin's *Gedankenexperiment* (1996) . . . . . 108
6.15 The Afshar Experiment (2003)* . . . . . 111
6.16 Momentum-Entangled 2-Slit Interference Experiments (1995–1999) . . . . . 113
6.16.1 The Ghost-Interference Experiment (1995)* . . . . . 113
6.16.2 The Dopfer Experiment (1999)* . . . . . 114
6.17 "Boxed Atom" Experiments (1992–2006) . . . . . 116
6.17.1 The Hardy One-Atom *Gedankenexperiment* . . . . . 116
6.17.2 The Elitzur–Dolev Three-Atom *Gedankenexperiment* . . . . . 117
6.17.3 The Elitzur–Dolev Two-Atom *Gedankenexperiment* . . . . . 120
6.17.4 The Time-Reversed EPR *Gedankenexperiment* . . . . . 121
6.17.5 The Quantum Liar Paradox . . . . . 124
6.18 The Leggett–Garg Inequality and "Quantum Realism" (2007)* . . . . . 124
6.19 Entanglement Swapping (1993–2009)* . . . . . 126
6.20 Gisin: Neither Sub- nor Superluminal "Influences"? (2012) . . . . . 128
6.21 The Black Hole Information Paradox (1975–2015) . . . . . 129

6.22 Paradox Overview. . . . . . . . . . 130
References. . . . . . . . . . 131

**7 Nonlocal Signaling?** . . . . . . . . . . 135
7.1 No-Signal Theorems . . . . . . . . . . 135
7.2 Nonlocal Signals and Special Relativity. . . . . . . . . . 136
7.3 Entanglement-Coherence Complementarity and Variable Entanglement . . . . . . . . . . 137
7.4 A Polarization-Entangled EPR Experiment with Variable Entanglement . . . . . . . . . . 138
7.5 A Path-Entangled EPR Experiment with Variable Entanglement* . . . . . . . . . . 139
7.6 A Wedge-Modified Path-Entangled EPR Experiment with Variable Entanglement . . . . . . . . . . 142
7.7 A Transactional Analysis of the Complementarity of One- and Two-Particle Interference . . . . . . . . . . 144
7.8 Singles Detection and the Absence of 1-Particle Interference . . . . . . . . . . 147
7.9 Entangled Paths and Hidden Signals: A Proof . . . . . . . . . . 148
7.10 Conclusions about Nonlocal Signals . . . . . . . . . . 149
References. . . . . . . . . . 150

**8 Quantum Communication, Encryption, Teleportation, and Computing**. . . . . . . . . . 151
8.1 Quantum Encryption and Communication and the TI . . . . . . . . . . 151
8.2 Quantum Teleportation and the TI . . . . . . . . . . 152
8.3 Quantum Computing and the TI . . . . . . . . . . 155
References. . . . . . . . . . 159

**9 The Nature and Structure of Time** . . . . . . . . . . 161
9.1 The Arrows of Time . . . . . . . . . . 161
9.2 Determinism and the TI. . . . . . . . . . 165
9.3 The Plane of the Present and the TI . . . . . . . . . . 165
References. . . . . . . . . . 166

**10 Conclusion** . . . . . . . . . . 167
Reference . . . . . . . . . . 168

**Appendix A: Frequently Asked Questions About Quantum Mechanics and the Transactional Interpretation** . . . . . . . . . . 169

**Appendix B: A Brief Overview of the Quantum Formalism** . . . . . . . . . . 189

**Appendix C: Quantum Dice and Poker—Nonlocal Games of Chance**. . . . . . . . . . 197

**Appendix D: Detailed Analyses of Selected *Gedankenexperiments*** .... 199

**Index** .... 213

**(Sections labeled with an asterisk (*) indicate real experiments that have been performed in the quantum optics laboratory.)**

# About the Author

**John G. Cramer** is Professor Emeritus, Physics, at the University of Washington (UW) in Seattle, where he has had five decades of experience in teaching undergraduate and graduate level physics. He has done cutting-edge research in experimental and theoretical nuclear and ultra-relativistic heavy ion physics, including active participation in Experiments NA35 and NA49 at CERN, Geneva, Switzerland, and the STAR Experiment at RHIC, Brookhaven National Laboratory, Long Island, NY. He has also worked in the foundations of quantum mechanics (QM) and is the originator of QM's Transactional Interpretation. He served as Director of the University of Washington Nuclear Physics Laboratory from 1983 to 1990, overseeing a major $10,000,000 accelerator construction project.

John has also served on accelerator-laboratory Program Advisory Committees for LAMPF (Los Alamos National Laboratory), NSCL (Michigan State University), TRIUMF (University of British Columbia), and the 88" Cyclotron (Lawrence Berkeley National Laboratory). He is a Fellow of the American Association for the Advancement of Science and of the American Physical Society (APS), was Chair of the APS/DNP Nuclear Science Resources Committee (1979–1982), and served on the APS Panel on Public Affairs (1998–2003). He presently serves on the External Council of the NIAC innovative-projects program of NASA.

John has spent three 15-month sabbaticals in Europe, the first (1971–1972) as Bundesministerium Gastprofessor, Ludwig-Maximillian-Universität-München, Garching, Germany; then (1982–1983) as Gastprofessor, Hahn-Meitner Institut, Berlin; and finally (1994–1995) as Guest Researcher, Max-Planck Institut für Physik, München, with three months of this sabbatical spent at CERN as Experiment NA49 came into operation. He is co-author of almost 300 publications in nuclear and ultra-relativistic heavy ion physics published in peer-reviewed physics journals, as well as over 141 publications in conference proceedings, and has written several chapters for multiauthor books about physics.

John is the author of the award-nominated hard science fiction novels **Twistor** and **Einstein's Bridge**, both published by Avon Books. **Twistor** is currently available as a Dover reprint and as an e-book from Book View Cafe. **Einstein's Bridge** will soon be joined by a new sequel, **Fermi's Question**, both to be published by Tor Books. John is also the author of over 181 popular-level science articles published bimonthly from 1984 to present in his "The Alternate View" columns appearing in every-other issue of **Analog Science Fiction and Fact Magazine**.

John was born in Houston, Texas on October 24, 1934, and was educated in the Houston Public Schools (Poe, Lanier, Lamar) and at Rice University, where he received a BA (1957), MA (1959), and Ph.D. (1961) in Experimental Nuclear Physics. He began his professional physics career as a Postdoc and then Assistant Professor at Indiana University, Bloomington, Indiana (1961–1964) before joining the Physics Faculty of the University of Washington. John and his wife Pauline live in the View Ridge neighborhood of Seattle, Washington, with their three Shetland Sheepdogs, MACH-4 Lancelot, MACH Viviane, and Taliesin.

# List of Figures


| | | |
|---|---|---|
| Figure 0.1 | Carver Mead (source: Carver Mead.) | viii |
| Figure 1.1 | Lawrence Biedenharn (source: Duke Univ.) | 2 |
| Figure 1.2 | Biedenharn's blackboard (source: JGC) | 2 |
| Figure 1.3 | NPL blackboard showing EPR V-diagram (source: JGC) | 5 |
| Figure 1.4 | Quantum beer bottle in transit (source: JGC) | 7 |
| Figure 2.1 | William Thompson, Lord Kelvin (source: public domain) | 10 |
| Figure 2.2 | Max Planck (source: public domain) | 11 |
| Figure 2.3 | Albert Einstein (source: public domain) | 11 |
| Figure 2.4 | J. J. Thompson (source: public domain) | 12 |
| Figure 2.5 | Ernest Rutherford (source: public domain) | 12 |
| Figure 2.6 | Niels Bohr (source: public domain) | 13 |
| Figure 2.7 | Louis de Broglie (source: public domain) | 14 |
| Figure 2.8 | Werner Heisenberg (source: public domain) | 15 |
| Figure 2.9 | Erwin Schrödinger (source: public domain) | 18 |
| Figure 2.10 | Gaussian pulse in time and frequency domains (source: JGC) | 21 |
| Figure 2.11 | Heisenberg's Microscope (source: JGC) | 23 |
| Figure 2.12 | Max Born (source: public domain) | 25 |
| Figure 2.13 | John Stuart Bell (source: CERN Archive) | 31 |
| Figure 2.14 | Malus' Law (source: JGC) | 32 |
| Figure 2.15 | EPR noise rate vs. angle (source: JGC) | 34 |
| Figure 2.16 | John Clauser (source: John Clauser) | 35 |
| Figure 2.17 | Stuart Freedman (source: LBNL *Currents*) | 36 |
| Figure 3.1 | Hans Kramers (source: public domain) | 41 |
| Figure 3.2 | Arthur Compton (source: public domain) | 42 |

Figure 4.1 Variable in the complex plane (source: JGC). . . . . . . . . . . . 48
Figure 4.2 Paul Dirac (source: public domain) . . . . . . . . . . . . . . . . . . 49
Figure 4.3 Richard Feynman (source: Jesse Jusek, artist) . . . . . . . . . . . 52
Figure 4.4 John Wheeler (source: University of Texas) . . . . . . . . . . . . 53
Figure 4.5 Wheeler-Feynman handshake (source: William R. Warren, Jr., artist) . . . . . . . . . . . . . . . . . . . . . . . . . . . . . . . . 54

Figure 5.1 Emission stage of 1D Transaction (source: JGC) . . . . . . . . . 60
Figure 5.2 Confirmation stage of 1D Transaction (source: JGC) . . . . . . 61
Figure 5.3 Completed 1D Transaction (source: JGC) . . . . . . . . . . . . . . 61
Figure 5.4 Offer wave $\psi$ (source: JGC) . . . . . . . . . . . . . . . . . . . . . . . . 62
Figure 5.5 Confirmation wave $\psi^*$ (source: JGC) . . . . . . . . . . . . . . . . 63
Figure 5.6 Completed transaction as standing wave (source: JGC). . . . . 64
Figure 5.7 Mead model of transaction formation (source: JGC). . . . . . . 67

Figure 6.1 Young's Two-Slit Experiment (source: JGC). . . . . . . . . . . . 76
Figure 6.2 Build-up of 2-slit Interference Pattern (source: Prof. Antoine Weis, Univ. of Fribourg, Switzerland, and Prof. Todorka Dimitrova, Plovdiv Univ., Bulgaria.) . . . . . . . . . . . . . . . . . . . . . . . . . . 77
Figure 6.3 Einstein's Bubble Paradox (source: JGC) . . . . . . . . . . . . . . 79
Figure 6.4 Schrödinger's Cat Paradox (source: William R. Warren, Jr., artist) . . . . . . . . . . . . . . . . . . . . . . . . . . . . . . . . 80
Figure 6.5 Wigner's Friend Paradox (source: William R. Warren, Jr., artist) . . . . . . . . . . . . . . . . . . . . . . . . . . . . . . . . 83
Figure 6.6 Renninger's *Gedankenexperiment* (source: JGC) . . . . . . . . . 84
Figure 6.7 Photon through polatizers (source: JGC). . . . . . . . . . . . . . . 86
Figure 6.8 Wheeler's Delayed Choice Experiment (source: JGC). . . . . . 89
Figure 6.9 Nonlocal "V" Transaction (source: JGC). . . . . . . . . . . . . . . 92
Figure 6.10 Hanbury-Brown Twiss Experiment (source: JGC) . . . . . . . . 93
Figure 6.11 AAD Predictions (source: JGC). . . . . . . . . . . . . . . . . . . . . 95
Figure 6.12 TI analysis of AAD (source: JGC). . . . . . . . . . . . . . . . . . . 97
Figure 6.13 Quantum Eraser Experiment (source: JGC) . . . . . . . . . . . . 99
Figure 6.14 Mach Zehnder Interferometer (source: JGC) . . . . . . . . . . . . 101
Figure 6.15 Mach Zehnder Interferometer with beam blocked (source: JGC) . . . . . . . . . . . . . . . . . . . . . . . . . . . . . . . . 102
Figure 6.16 Offer waves–1 (source: JGC) . . . . . . . . . . . . . . . . . . . . . 103
Figure 6.17 Offer waves–2 (source: JGC) . . . . . . . . . . . . . . . . . . . . . 104
Figure 6.18 Quantum Zeno Experiment (source: JGC). . . . . . . . . . . . . . 105
Figure 6.19 Quantum Zeno Experiment unfolded (source: JGC) . . . . . . . 106
Figure 6.20 Blocked Quantum Zeno Experiment (source: JGC). . . . . . . . 107
Figure 6.21 Quantum Zeno Experiment confirmation waves (source: JGC) . . . . . . . . . . . . . . . . . . . . . . . . . . . . . . . . 107
Figure 6.22 Maudlin's *gedankenexperiment* (source: JGC) . . . . . . . . . . 109

Figure 6.23 Afshar Experiment (source: JGC) . . . . 111
Figure 6.24 Ghost Interference Experiment (source: JGC) . . . . 113
Figure 6.25 Dopfer Experiment (source: JGC) . . . . 115
Figure 6.26 Hardy Single-Atom Experiment (source: JGC) . . . . 116
Figure 6.27 Elitzur-Dolev 3-Atom Experiment (source: JGC) . . . . 118
Figure 6.28 Elitzur-Dolev 2-Atom Experiment (source: JGC) . . . . 120
Figure 6.29 Time-Reversed EPR Experiment (source: JGC) . . . . 122
Figure 6.30 Quantum Liar Paradox (source: JGC) . . . . 123
Figure 6.31 Entanglement Swapping (source: JGC) . . . . 127

Figure 7.1 Polarization-entangled EPR Experiment (source: JGC) . . . . 138
Figure 7.2 Path-entangled EPR Experiment (source: JGC) . . . . 141
Figure 7.3 Wedge Path-entangled EPR Experiment (source: JGC) . . . . 142
Figure 7.4 Transaction linking detectors and source–1 (source: JGC) . . . . 145
Figure 7.5 Transaction linking detectors and source–2 (source: JGC) . . . . 145
Figure 7.6 Transaction linking detectors and source–3 (source: JGC) . . . . 146
Figure 7.7 Transaction linking detectors and source–4 (source: JGC) . . . . 146

Figure 8.1 Functionalized Porphyrin (source: Prof. M. Arndt) . . . . 155

Figure 9.1 Alternative time-arrow hierarchies (source: JGC) . . . . 162
Figure 9.2 Ludwig Boltzmann (source: public domain) . . . . 163

Figure B.1 Localized wave function (source: JGC) . . . . 191
Figure B.2 Vertically-polarized light wave (source: JGC) . . . . 192

Figure D.1 Hardy Single-Atom Experiment (source: JGC) . . . . 200
Figure D.2 Offer Waves (source: JGC) . . . . 201
Figure D.3 Polarization-entangled EPR Experiment (source: JGC) . . . . 203
Figure D.4 Joint detection probabilities vs. $\theta$ (source: JGC) . . . . 204
Figure D.5 Path-entangled EPR Experiment (source: JGC) . . . . 205
Figure D.6 Bob's singles detector probabilities (source: JGC) . . . . 206
Figure D.7 Wedge path-entangled EPR Experiment (source: JGC) . . . . 208
Figure D.8 Wave function magnitudes (source: JGC) . . . . 208
Figure D.9 Probabilities for coincident detection (source: JGC) . . . . 209
Figure D.10 Difference plot showing no nonlocal signal (source: JGC) . . . . 210

# Chapter 1
# Introduction

## 1.1 My Adventures with Advanced Waves and Time-Symmetry

When I was a graduate student in 1958, I had a disagreement with my E & M teacher, Associate Professor Larry Biedenharn (Fig. 1.1), the then-young hot-shot nuclear theorist who that year was teaching electromagnetic theory to us beginning physics graduate students at Rice University. In his lecture, he had just gone through the manipulation of Maxwell's equations to produce the electromagnetic wave equation. He pointed out that the wave equation, because it was second-order in time, had two independent time solutions. He wrote both solutions on the blackboard and then drew a big X through one of them, the advanced-wave solution (Fig. 1.2). He said that this was because the advanced solution went in the wrong time direction and violated the "causality boundary condition". The other one, the retarded-wave solution, was OK, because it was consistent with causality and the idea that a time delay was required for light to propagate from one location to another.[1]

I had previously taken some rather demanding mathematics courses at Rice that focused on functions of a complex variable and on solutions of a variety of differential equations. Based on that background, I raised my hand and challenged the notion that *causality*, the idea that a cause must precede its effects in time sequence, could be used as a boundary condition, as he had just done. "Boundary conditions describe known conditions at boundaries," I said dogmatically. We argued this point back and forth for several minutes. Finally, he told me to see him after class, and he continued with the lecture on the use of retarded waves and retarded potentials, based on causality considerations, in solving various radiation and electrodynamics problems.

[1]The advanced solution of the wave equation is called "advanced" because the wave arrives *before* it departs, and time advances the wave's phase; the more ordinary retarded solution arrives *after* it departs, and time retards the wave's phase.



J.G. Cramer, *The Quantum Handshake*, DOI 10.1007/978-3-319-24642-0_1

**Fig. 1.1** Lawrence C. Biedenharn (1922–1996), my E & M teacher

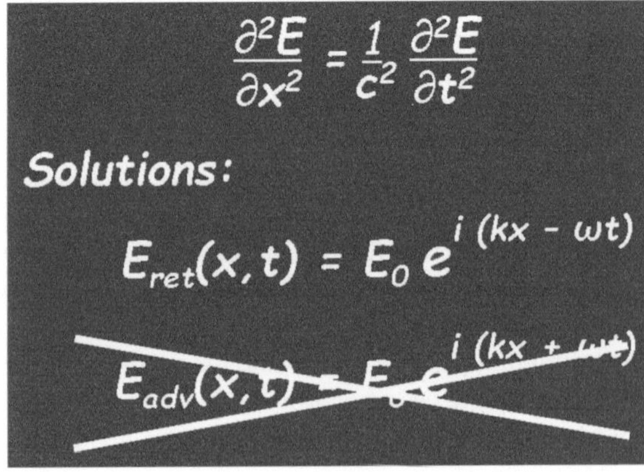


**Fig. 1.2** Biedenharn's blackboard with the advanced wave solution of the electromagnetic wave equation crossed out

After class, Biedenharn told me that I had interesting ideas, but they were wrong. He said that two theorists of his acquaintance, Johnny Wheeler and Dick Feynman, had previously explored the point I was making by assuming that one should be even-handed in the treatment of time in electrodynamics and should use a symmetric mixture of advanced and retarded solutions, thereby avoiding building an "arrow of time" into electrodynamics by the arbitrary selection of retarded waves and potentials. He said that I should go to the library and look in the 1945 issues of the journal *Reviews of Modern Physics* for their article [1]. He said that they had done exactly what I wanted, but it had its own problems and was not useful for doing electrodynamics calculations. Later that day I duly went to Rice's Fondren Library, found the appropriate issue of *Reviews of Modern Physics*, and read it carefully.

Larry Biedenharn was right. Wheeler and Feynman had constructed a time-symmetric formalism and then used real boundary conditions involving future absorbers to justify causality and the arrow of time that we observe in the real world. Retarded waves spread out from an emitter in the present and were answered by advanced waves from absorbers in the future. A "handshake" between retarded and advanced waves arranged for the transfer of energy and momentum and accounted for the recoiling of the emitter. The absorber's electric charge responded to the retarded field and the emitter's charge responded to the advanced field. This approach to electrodynamics satisfied my mathematical intuitions about how differential equations should be solved and solutions selected.

However, in practice it looked hard to do calculations with the formalism, because one had to integrate over the entire future universe. Further, the Wheeler–Feynman assertion about the dominance of future over past absorbers looked fishy to me. Wasn't the universe supposed to be expanding and growing less dense as time progressed? Why would there be more absorption in the diffuse future than in the dense past? I concluded that it was an interesting paper, but that it was not the way to do electrodynamics, which I was supposed to be learning. And so I put aside Wheeler and Feynman and focused on learning to become a physicist. But it was in this way that Larry Biedenharn set me on the path that would eventually lead to the Transactional Interpretation of quantum mechanics.

I had started graduate work at Rice in 1957, the year that T. D. Lee and C. N. Yang received the Nobel Prize in Physics for their discovery that parity was not conserved in weak interaction processes like beta-decay [2]. My Master's thesis at Rice was an experimental demonstration, detecting circularly polarized *bremsstrahlung*, of the non-conservation of parity in the beta decay of $^{8}$Li, a nucleus that was short-lived and had a decay energy about 10 times higher than the decay used as the initial experimental demonstrations of parity non-conservation.[2] The work for my PhD thesis started in 1959 and was an experimental test of conserved vector current, as predicted by Murray Gell-Mann, in a detailed comparison of the $^{12}$N $\beta^{+}$ and $^{12}$B $\beta^{-}$ beta-decay spectra that I measured.

In 1961, I finished my dissertation and went on to a Postdoc position at Indiana University, where I studied nuclear reactions and angular correlations using the 22 MeV alpha-particle beam of the old Indiana University cyclotron, built by Milo Sampson in the 1950s. After three years and some very productive work at Indiana, I moved to the University of Washington as an Assistant Professor. The UW Nuclear

[2]My Masters Thesis observing parity non-conservation in the $^{8}$Li decay won the Rice Best Thesis Award for 1959.

Physics Laboratory was getting a brand new three-stage FN tandem Van de Graaff accelerator and a state-of-the-art SDS 930 online data-collection computer that I wanted to play with.

My work at UW used lots of quantum mechanics as applied to nuclear reactions, but it took me very far from any worries about its interpretation. I developed some skill in numerically solving the Schrödinger and Klein-Gordon equations on a computer and in using these numerical wave-function solutions to predict the nuclear scattering and reaction cross sections that we were measuring in the laboratory. I even spent some time developing a computer-based nonlocal nuclear optical model program to study the possibility that nonlocal potentials were acting in heavy ion scattering [3]. I should emphasize that we never thought of the quantum wave functions we were calculating as representing the knowledge of some hypothetical observer or residing in some abstract Hilbert space. We always thought of them as representing real particle-waves moving through space, interacting, and colliding. This experience at the boundary between experiment and theory gave me a "nuts-and-bolts" appreciation of and intuition about quantum processes that few abstraction-prone quantum theorists or philosophers possess.

In the late 1970s, some twenty years after my initial exposure to Wheeler and Feynman, I was teaching modern physics[3] to a class of undergraduates at the University of Washington. We had been discussing the peculiarities and paradoxes of quantum mechanics, and I was describing the "Einstein's Bubble paradox" (see Sect. 6.2), the idea that the quantum wave function represented by the symbol $\psi$ must somehow know when it was time to disappear from a region in which it had been spreading, at the instant when the particle it described was being detected elsewhere.

Suddenly and unexpectedly, in a flash of insight that came in mid-sentence, I realized that I understood a way in which quantum mechanics could do that. There were retarded waves ($\psi$) and advanced waves ($\psi^*$) all over the quantum formalism. If these advanced and retarded waves were having Wheeler–Feynman handshakes, the behavior that worried Einstein could be easily explained. Only those retarded waves that made advanced-wave echoes could do particle transfers. Einstein's paradox was solved without the need of invoking Heisenberg's highly questionable knowledge interpretation. I smiled, filed this insight away for future attention, and continued with my lecture.

A month or so after that, I was in the coffee room of the University of Washington Nuclear Physics Laboratory (NPL), where I did most of my physics research. Rick Norman, then a postdoc at the NPL and now a distinguished senior researcher at the Lawrence Berkeley National Laboratory, was talking about some of the recent Einstein–Podolsky–Rosen experiments that were testing Bell's inequalities and demonstrating experimentally that hidden variable theories were wrong and that quantum mechanics was nonlocal. He had been thinking about possibly doing an

[3]The term "modern physics" is an anachronistic colloquialism of the physics community, meaning special relativity, introductory quantum mechanics, and atomic and nuclear structure and behavior. The term refers to the post-classical physics that was "modern" when it was developed in the early 20th century, almost 100 years ago.

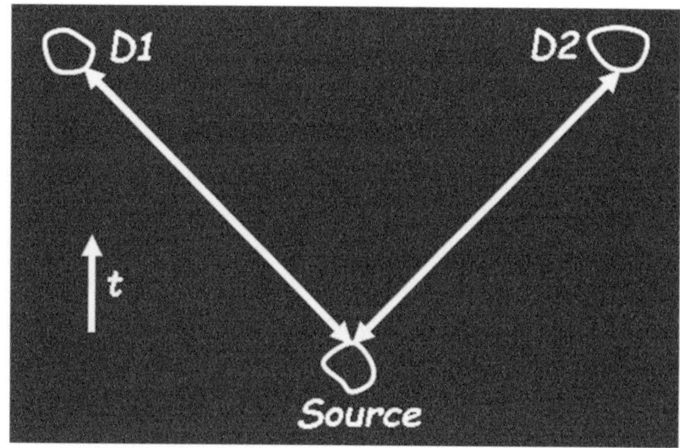


**Fig. 1.3** NPL blackboard showing the "V" diagram of EPR handshakes

EPR experiment and was wondering whether there was anything that we could do with the facilities we had available at the NPL to do some testing and pushing of the envelope. One could perhaps mount accelerator-based experiments with polarized protons or detect gamma ray polarization by scattering (see Sect. 2.7). However, both types of experiment had many more experimental difficulties than the relatively simple optical EPR experiments that were presently being done with polarized light.

I commented that I had developed a way of thinking about EPR experiments and that I understood how nonlocality worked. Rick looked surprised and asked me to explain. On the blackboard (Fig. 1.3), I drew a V-shaped diagram and explained how the dual Wheeler–Feynman handshakes connecting the entangling 2-photon source to the two EPR polarization detectors must match at both detectors and back at the source so that angular momentum is conserved, producing the needed correlations. Rick asked if I had published this. "No," I said, "I thought it was obvious." It was then that I realized that perhaps it wasn't so obvious.

Over the next months, I scoured the physics literature to see if the Wheeler–Feynman ideas had ever been applied to the interpretation of quantum mechanics. The answer seemed to be *no*. Louis de Broglie [4] had proposed a "pilot-wave" model, with trail-blazing quantum waves showing the trailing particles which path to take, that looked a lot like the initial stage of a Wheeler–Feynman handshake, but it stopped there, without the time-reversed confirmation. There had been some work by Hoyle and Narlikar [5, 6] and by Paul Davies [7–9], independently producing Wheeler–Feynman-inspired time-symmetric versions of quantum electrodynamics. Olivier Costa de Beauregard, a former student of de Broglie in France, had attempted to account for quantum nonlocality in terms of Feynman's zig-zag diagrams [10–17]. But interpreting the psi-stars of standard quantum mechanics as Wheeler–Feynman advanced waves and as participants in handshakes across space-time seemed to be completely new ideas.

I was (and am) an experimental nuclear physicist, not a philosopher of science or an abstract theoretician. I was skilled in making hardware and accelerators work and in using computers for data collection and for performing numerical quantum mechanics calculations. There is a saying that "physics is what physicists do late at night", and that certainly applies to experimental physics. As an experimentalist, I have found great reward in stepping outside the accelerator laboratory in the middle of the night, looking up at the stars, and realizing that I was the only person on the planet, and perhaps in the entire galaxy, that knew some particular thing about the way the universe works, because my experimental research had just revealed it to me in the laboratory that night.

I always grew uncomfortable when physics concepts became so abstract that I could not visualize them, but I was a really excellent visualizer. So I adopted a Zen-like "beginner's mind" approach to the quantum puzzle. I attempted to look within the inscrutable quantum formalism itself from the Wheeler–Feynman point of view, to peer into the formalism and visualize the inner mechanisms that produced the mathematics. This led to two papers discussing aspects of the approach [18, 19], and culminated several years later with my long 1986 paper [20], written mainly while I was on sabbatical at the Hahn-Meitner Institute in West Berlin. The paper was published in the journal *Reviews of Modern Physics*, some 41 years after Wheeler and Feynman published in the same journal, and it introduced the Transactional Interpretation of quantum mechanics.

That was almost 30 years ago, and while the Transactional Interpretation had received considerable attention, including the science popularization, *Schrödinger's Kittens* by John Gribbin [21], the prize-winning play *now, then, again* by Penny Penniston [22], who used a romantic plot-line based on the Transactional Interpretation, and the philosophy-oriented book, *The Transactional Interpretation of Quantum Mechanics*[4] by Ruth Kastner [23], it has only managed to join the ranks of the large number of alternative interpretations of quantum mechanics. Its virtues of simplicity, comprehensiveness, explanation power, and visualization qualities have not been widely appreciated, and it has not received the general acceptance that I had expected.

This book attempts to provide a new comprehensive introduction to the Transactional Interpretation, as an alternative to the orthodox Copenhagen Interpretation, and to show that it solves *all* of the interpretational problems and paradoxes of quantum mechanics and, in particular, shines a very bright light on the inner mechanisms behind quantum nonlocality. We will begin by looking closely at quantum mechanics as it presently exists.

## 1.2 Quantum Mechanics and Beer Bottles

Quantum mechanics is the most unusual of scientific theories. For readers struggling to appreciate the science, I need to explain. In the history of physics almost all new theories, Newton's mechanics, Maxwell's electromagnetism, thermodynamics, Einstein's special relativity, have been developed in more or less the same way. The

[4]We note that despite its title, Kastner's book introduces the *possibilist* transactional interpretation, a variant of the Transactional Interpretation presented here that attempts to explain quantum nonlocality in a qualitatively different way.

**Fig. 1.4** Quantum wave-particle beer bottle in transit

originator had some basic concept, some picture of the way the universe worked in the area of interest. Then he carefully constructed the mathematical formalism that converted that picture into a theory capable of making testable predictions that could be compared with the results of observations, experiments, and measurements. It has always worked like that ..., except in the case of quantum mechanics.

Quantum mechanics is our current standard theory for describing energy and matter at the smallest scales and for predicting the behavior of fundamental particles and composite systems like atoms and molecules. As we will see in Chap. 2, the formalism of quantum mechanics did not originate in the same way as other theories of physics. It was produced by Heisenberg and Schrödinger in two quite different ways, but with the ultimate result that there was no accompanying picture of the inner workings of the systems described. This would be less of a problem if the behavior of objects like electrons, as described by quantum mechanics, was simple and straightforward, but in the quantum world that behavior is quite bizarre (Fig. 1.4).

> As a fanciful example, consider the following analogy. Suppose that I were in Boston, and I went down to the docks and threw a beer bottle into the waters of Boston Harbor. The bottle disappeared, producing waves that spread out across the Atlantic Ocean in all directions from the entry point. Some of these waves traveled toward England and France and Spain and Western Africa and Eastern South America. In particular, some waves from the bottle traveled over the North Sea to the harbor area of Copenhagen. There the waves abruptly disappeared, and my beer bottle suddenly appeared in its original form on the Copenhagen dock. At the instant of its appearance, the waves traveling elsewhere around the Atlantic abruptly disappeared too.[5] In this scenario, my beer bottle has moved quantum mechanically from Boston to Copenhagen.

Clearly beer bottles do not behave in this way, but this just is how quantum mechanics describes the movement of an electron or a photon from one location to another. As rational beings who want to understand the world, we would like to make sense of such behavior. However, today's standard model for the microscopic world, accepted by most physicists, is a combination of the inscrutable mathematical formalism of quantum mechanics and the Copenhagen Interpretation of that formalism.

---

[5] At least, in the Copenhagen view, the waves would abruptly disappear. From the viewpoint of the Transactional Interpretation, presented here, they would keep on going to other locations and send confirmation waves back to Boston.

The Copenhagen Interpretation provides a very useful guide for how to use the formalism to make predictions that can be compared with experiment and observation, but it tells us little else.

The Copenhagen Interpretation is sometimes described as the "don't ask; don't tell" interpretation of the quantum world. It advises us to focus on measurable quantities and not to inquire about what is going on behind the scenes, to "shut up and calculate." This view derives in part from the philosophy of logical positivism, which was fashionable in the philosophical circles of Berlin and Vienna in the late 1920s, at the time when quantum mechanics was being developed. Philosophy has moved well beyond logical positivism, but the Copenhagen Interpretation remains stuck in that era, like a carboniferous insect encased in amber. The Copenhagen Interpretation is a consistent but unsatisfying way of approaching the quantum world. In this book, we will provide a better alternative: *the Transactional Interpretation of quantum mechanics*.

## References

1. J.A. Wheeler, R.P. Feynman, Rev. Mod. Phy. **17**, 157 (1945)
2. T.D. Lee, C.N. Yang, Phys. Rev. **105**, 1671 (1957)
3. J.G. Cramer, R.M. DeVries, Effects of non-local potentials in heavy ion reactions. Phys. Rev. C **14**, 122–126 (1976)
4. L. de Broglie, J. de Physique et du Radium **8**, 225 (1927)
5. F. Hoyle, J.V. Narlikar, Ann. Phys. **54**, 207 (1969)
6. F. Hoyle, J.V. Narlikar, Ann. Phys. **62**, 44 (1971)
7. P.C.W. Davies, Proc. Camb. Philos. Soc. **68**, 751 (1970)
8. P.C.W. Davies, J. Phys. A **4**, 836 (1971)
9. P.C.W. Davies, J. Phys. A **5**, 1025 (1972)
10. O. Costa de Beauregard, C. R. Acad. Sci. Paris **236**, 1632 (1953)
11. O. Costa de Beauregard, Dialectica **19**, 280 (1965)
12. O. Costa de Beauregard, C. R. Acad. Sci. Paris **282**, 1251 (1976)
13. O. Costa de Beauregard, Phys. Lett. **60A**, 93 (1977)
14. O. Costa de Beauregard, Nuovo Cimento **42B**, 41 (1977)
15. O. Costa de Beauregard, Phys. Lett. **67A**, 171 (1978)
16. O. Costa de Beauregard, Lett. Nuovo Cimento **26**, 135 (1979)
17. O. Costa de Beauregard, Nuovo Cimento **51B**, 267 (1979)
18. J.G. Cramer, Phys. Rev. D **22**, 362–376 (1980)
19. J.G. Cramer, Found. Phys. **13**, 887 (1983)
20. J.G. Cramer, The transactional interpretation of quantum mechanics. Rev. Mod. Phys. **58**, 647–687 (1986)
21. J. Gribbin, *Schrödinger's Kittens and the Search for Reality: Solving the Quantum Mysteries* (Little Brown & Co., Boston, 1995). ISBN: 978-0316328388
22. P. Penniston, *Now Then Again*, (Broadway Play Pub., 2014). ISBN: 978-0881456028
23. R.E. Kastner, *The Transactional Interpretation of Quantum Mechanics: The Reality of Possibility* (Cambridge University Press, Cambridge, 2012)

# Chapter 2
# The Curious History of Quantum Mechanics

We start with an account of the history of physics in the early 20th century as it applies to the development of quantum mechanics and its interpretation. This account is intended to show how the theory and interpretation were developed and came to be in the state in which we find them today.

Writing about the development of science can be frustrating to the writer, because there are so many people involved, so many false starts, so many mistaken ideas, so much bad data, and a tangled story-line of rejected ideas and falsified theories left behind. I will attempt to streamline the lines of development by focusing on the ideas that turned out to be better and more important. I will, for the most part, ignore the bad ideas, the misconceptions, and the minor contributions. This perhaps distorts history, but I am not a historian, and I feel no obligation to achieve historical precision. Rather, I want to communicate the feel of the intellectually turbulent times when quantum mechanics was emerging. Many individual contributions will be ignored or neglected, but a picture of the development of quantum mechanics, with all its triumphs, paradoxes, and problems, should emerge.

## 2.1 Atomic Theory in the Early 20th Century (1900–1924)

William Thompson, Lord Kelvin (Fig. 2.1), was one of the founders of thermodynamics and perhaps the most prominent British scientist of his day. In 1900, he is reputed to have told the British Association for the Advancement of Science that there was nothing new to be discovered in physics and that all that remained to do was to make more and more precise measurements. One of his contemporaries, confident of the universality of Newtonian mechanics, similarly asserted that if one knew the positions and velocities of all the particles in the universe at one particular moment, the entire past and future of the universe could be calculated. This confidence in the finality of the physics of the time prevailed, despite the presence of certain unexplained loose ends. For example, in 1900 there were no good explanations of the line structure of atomic spectra, the energy source that powered stars, or



J.G. Cramer, *The Quantum Handshake*, DOI 10.1007/978-3-319-24642-0_2

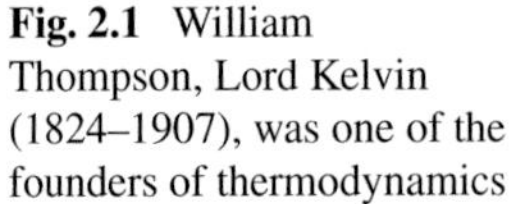

**Fig. 2.1** William Thompson, Lord Kelvin (1824–1907), was one of the founders of thermodynamics

the nature of radioactivity. These unexplained "details" were clouds on the horizon that foreshadowed a great intellectual storm, a scientific revolution in the making.

The physics of the 19th century had firmly established that light was an electromagnetic wave. Thomas Young's two-slit experiment (see Sect. 6.1) had demonstrated in the early 1800s that light waves taking two paths to reach a screen could be made to interfere, canceling at some locations on a screen and reinforcing in others. In the 1860s James Clerk Maxwell, starting with a set of equations governing the behavior of stationary or slowly changing electric and magnetic fields, had derived a wave equation that described self-sustaining coupled electric and magnetic waves moving through space at the speed of light.

At the beginning of the new 20th century, this widely accepted wave picture of light was jarred by the work of Max Planck (Fig. 2.2) in 1901. Planck showed that heated objects could only emit light in "energy chunks" of an energy size given by the frequency $f$ of the light multiplied by a new physical constant, which we now call Planck's constant and denote by the symbol $h$ (see Appendix A.1).

Planck interpreted his results as demonstrating the peculiarities of the emitting system and insisted that they were *not* describing an intrinsic property of light itself. In 1905, however, this view was challenged by Albert Einstein (Fig. 2.3), who showed that the photoelectric effect, the emission of electrons from metals illuminated by light, can be consistently explained by assuming that light itself has particle-like properties, with each particle (or photon) of light carrying an energy $E_\gamma$ equal to its frequency $f$ multiplied by Planck's constant $h$ (i.e., $E_\gamma = hf = \hbar\omega$, where $\hbar = h/2\pi$ and angular frequency $\omega = 2\pi f$).

**Fig. 2.2** In 1918, Max Planck (1858–1947) received the Nobel Prize in Physics for his work on black-body radiation

**Fig. 2.3** In 1921, Albert Einstein (1879–1955) received the Nobel Prize in Physics for his work on the photoelectric effect

The structure and behavior of the atom proved to be particularly vexing problems for the physicists of the early 20th century. J. J. Thompson (Fig. 2.4) discovered the negatively charged electron, a fundamental particle that somehow was a part of atoms. Ernest Rutherford (Fig. 2.5) demonstrated that the mass and the positive electric charge of atoms were both concentrated in a small central region (the atomic nucleus) much smaller than the size of the atom. These discoveries led to new insights.

**Fig. 2.4** In 1906, J. J. Thompson (1856–1940) received the Nobel Prize in Physics for the discovery of the electron

**Fig. 2.5** In 1908, Ernest Rutherford (1871–1937) received the Nobel Prize in Chemistry for the discovery of the atomic nucleus

Rutherford had suggested that each atom might be a tiny solar system, with the negatively-charged electron "planets" orbiting a central positive nuclear "sun". However, electrons in such paths would be continually changing direction with large accelerations, and the accepted electromagnetic theory of Maxwell required that such accelerated charges must produce light waves and must radiate away their energy and angular momentum in microseconds. But instead of continuous light radiation from unstable atoms, experimentalists observed that atoms in electrical discharges produced light only at specific narrow frequencies or "spectral lines" and that atoms were otherwise stable.

In 1913 Niels Bohr (Fig. 2.6) solved a part of the problem by placing constraints on Rutherford's solar-system model of the atom. Bohr's model allowed electrons to

**Fig. 2.6** In 1922, Niels Bohr (1885–1962) received the Nobel Prize in Physics for his atom model

orbit only in paths that had integer multiples of an angular momentum "quantum" given by Planck's constant $h$ divided by $2\pi$ (which physicists now denote by the symbol h-bar or $\hbar$). To change from one such atomic orbit to another, an electron had to make a "quantum jump", disappearing from one orbit and appearing in the other while changing energy and angular momentum and emitting a light photon that made up the energy difference. Bohr's model accounted for the stability of atoms (the electron orbits were stable states that did not radiate) and for spectral lines (as the photons produced in the well-defined quantum jumps), and it worked very well in explaining the structure and light radiated from the hydrogen atom, which consisted of a single electron orbiting a proton nucleus. However, when an atom had two or more electrons present, Bohr's model failed miserably. Bohr's model worked only for hydrogen and for hydrogen-like ionized atoms with only one orbiting electron. It was telling a part of the story, but important pieces of the puzzle were still missing.

In his 1924 PhD thesis, the French nobleman Prince Louis de Broglie (Fig. 2.7) supplied another missing piece of the puzzle. He reasoned that since light had particle-like behavior, as shown by Einstein's analysis of the photoelectric effect, it was plausible that matter particles like electrons might show an analogous wave-like behavior. The wavelength $\lambda$ of a photon can be calculated by dividing Planck's constant $h$ by its momentum $p$ ($\lambda = h/p$). If electrons showed wave-like behavior, de Broglie reasoned, they might have wavelengths given by the same relation. He applied his wavelength relation to Bohr's model of the hydrogen atom, and he found that a precisely integer number of electron wavelengths, as calculated from his formula, fitted into the circumference of each stable orbit of Bohr's model as "standing waves". In other words, Bohr's assumption that each orbit had an integer number of $\hbar$ units of angular momentum was completely equivalent to assuming that

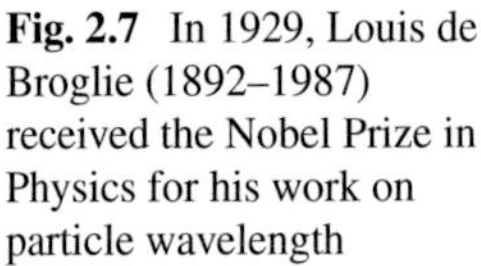

**Fig. 2.7** In 1929, Louis de Broglie (1892–1987) received the Nobel Prize in Physics for his work on particle wavelength

an integer number of de Broglie wavelengths of each electron fitted into its orbit. Each atomic electron is a stable "particle in a box" standing wave, with the box consisting of the electron's path bent into a closed circle or ellipse by the electric field of the nucleus. Later experimental work by Davisson and Germer in 1927 verified the concept by demonstrating that electrons could be made to show wave interference effects characteristic of their de Broglie wavelength when scattered from a crystal of nickel.

In the early 1920s, the stage was thus set for the development of a comprehensive theory of atomic structure and behavior, i.e., quantum mechanics. Some of the pieces of the puzzle had been provided by Einstein, Bohr, and de Broglie, while many others remained hidden. It would require at least two more major breakthroughs before the full theory of quantum mechanics, with all its power and peculiarities, could be realized.

## 2.2 Heisenberg and Matrix Mechanics (1925)

In late 1924, young Werner Heisenberg (Fig. 2.8) found that nothing seemed to make sense at Niels Bohr's Institute. The grateful Danish government, with the financial support of the Carlsberg Brewery, had provided their new Nobel Laureate with an endowed Institute for Theoretical Physics housed in a three-story building just outside Copenhagen. Here Bohr had gathered some of the world's brightest young theoretical physicists, including Hans Kramers, Wolfgang Pauli, and Werner Heisenberg, in an attempt to make sense of the rich data that were coming from the spectral lines of light emitted by excited atoms. They sought a way of generalizing the Bohr model so that it worked for all atoms instead of just for hydrogen.

**Fig. 2.8** In 1932, Werner Heisenberg (1901–1976) received the Nobel Prize in Physics for his work on matrix mechanics

The data from studies of atoms in electrical discharges showed a bewildering array of spectral lines that changed dramatically from one element to the next. There were mysterious double values or "doublets" in some lines, and there were strange shifts and splittings that depended on electric and magnetic fields. Clearly, Nature was trying to send an important message about the way the universe worked, but Bohr and his best and brightest had so far been unable to decode it. Starting from Bohr's application of angular momentum quantization to the hydrogen atom and his more recent work with John C. Slater and Hans Kramers attempting to ignore photons and use "virtual oscillators", they had tried model after model that would have built on Bohr's initial success. However, all attempts to picture what might be going on inside the atom, and then to develop mathematics appropriate to that picture, had utterly failed.

By May of 1925, Werner Heisenberg had moved from Copenhagen back to Max Born's Institute in Göttingen and was feeling burned out. He had been modeling an atom as a not-quite-perfect mass-and-charge system, a little anharmonic oscillator that had exotic Bessel functions instead of sine waves as its vibration modes. He had generated reams and reams of math, to no avail. The resulting atom orbits made frequencies that looked nothing like the light from real atoms. Ugly experimental reality had killed another lovely theoretical idea. The pictures at the heart of the calculations were somehow wrong.

In desperation, Heisenberg tried another approach, working directly with "laundry lists" of values describing the frequencies and strengths of atomic transitions and focusing on hydrogen, the simplest of the atoms. Somehow, these concrete variables, based on direct measurements, seemed to have more meaning as objects of theoretical significance than did the more ephemeral "unseen" variables that were implicit in the pictures and models behind the calculations. Heisenberg was reaching the conclusion that one should perhaps dismiss models altogether and focus exclusively on relationships between observable quantities.

But just when this new approach seemed to be on the verge of making progress, disaster struck. Heisenberg had always had problems with allergies, and every year the spring hay fever season had been a time of particular difficulty. Further, this year was much worse than usual. There had been a warm winter, and the pollen-producing vegetation of Northern Europe had outdone itself in the late spring. Heisenberg, attempting to work with Max Born in Göttingen, had been laid low with a bout of hay fever that was worse than anything in his previous 23 years. He could not sleep, and he walked around with swollen half-closed eyes, feeling as if his head was trapped inside a specimen bottle. He was unable to concentrate on anything.

A friend recommended Helgoland, a barren grassless island off the northern coast of Germany. The air there was the purest in Europe, there were few pollen-producing plants, and the growing season occurred several months later due to the colder climate of the North Sea. Heisenberg was now ensconced for ten days on the 2nd floor of a cozy guest cottage, with a nice view of the southern coast of the island. He was alone with the barren boulders, the pure air, newly cleared sinuses, a book by Goethe, and his lists and tables of observables.

*And then a miracle happened.* In later life, Werner Heisenberg was never able to adequately describe the mental processes that led to his breakthrough. Somehow, he had discovered a "new math", a systematic set of procedures that allowed him to manipulate the lists of numbers that were the focus of his inquiry into new lists that predicted other experimental observations. The new lists agreed with known data from the atomic spectroscopy of hydrogen. But Heisenberg's procedures were peculiar. Among other oddities, they violated the usual commutation rule of mathematics. Multiplying $P$ by $Q$ gave a different result from multiplying $Q$ by $P$.

Heisenberg wrote what he considered to be a "crazy" paper describing his new arcane procedures for producing new experimental results from other experimental results, and he pondered what to do with it. On his return to Göttingen, he gave a copy of the paper to his older friend and sometime employer, Max Born. Born and his mathematically skilled assistant, Pascal Jordan, immediately recognized the procedures Heisenberg had described as the manipulation of matrices. Heisenberg was completely unfamiliar with the mathematics of matrices, but nevertheless he had somehow invented it to fill the needs of his calculations. By November 1925, Heisenberg, Born, and Jordan had produced the matrix formulation of quantum mechanics, a powerful formal approach to making quantum mechanical predictions [1–3] that is still widely used in atomic and nuclear physics, particularly in shell-model calculations.

This new matrix approach was abstract (and nearly incomprehensible to the technically uninitiated). There was no underlying model to illustrate or justify the procedures used. Complex algebra was invoked, so that elements of matrices had both real and imaginary parts. Every variable and function of conventional atomic physics had to be reinterpreted as a matrix, a one- or two-dimensional list of values that was treated as a single object of manipulation. Continuous variables became infinite matrices. These elements were combined using matrix operations (addition, multiplication, inversion, diagonalization, extraction of eigenvalues, etc.) to yield “matrix elements” that were squared to make them real instead of complex and to obtain predictions for experimental observations.

The “why” of this matrix formalism was not apparent. The recommended approach, grounded in the logical positivism that was philosophically fashionable in the 1920s, was to “shut up and calculate”. Heisenberg had much accumulated frustration from the focus on pictures and models at Bohr’s Institute. He now found it rather liberating to reject such ephemeral pictures and models and to focus exclusively on experimental observations. And *matrix mechanics*, one flavor of the new standard theory of quantum mechanics, had emerged.

## 2.3 Schrödinger and Wave Mechanics (1926)

Erwin Schrödinger (Fig. 2.9) was a physicist on the move in post-WWI Europe. In early 1920, in rapid succession, he married Annemarie (Anny) Bertel and became an Assistant to Max Wien in Jena. Then in September, 1920, he became an “AO” (or associate) Professor in Stuttgart. Shortly afterwards, in early 1921, he became an “O” (or full) Professor in Breslau. Later in 1921 he moved to the University of Zurich and became an “O” Professor there. He stayed in Zurich for the next six years, and it was there that he made his breakthrough discovery.

In September, 1925 Schrödinger had obtained a copy of the 1924 French PhD thesis of Louis de Broglie, which outlined de Broglie’s matter-wave hypothesis and showed that treating electrons as orbital standing waves led to the angular momentum quantization constraint that was at the heart of Bohr’s model. Peter Debye, a Professor at Zurich’s ETH, persuaded Schrödinger to give a joint colloquium for their two institutions describing de Broglie’s work. At the end of this colloquium, which reportedly was a very clear and thoughtful presentation, Debye casually remarked that he considered de Broglie’s way of discussing waves to be rather naïve. He had learned as a student of Arnold Sommerfeld in Munich, he said, that to properly deal with waves, one must have a wave equation. But no wave equation was apparent in de Broglie’s work (Fig. 2.7).

Schrödinger puzzled over Debye’s remark while on a ski vacation in Arosa, a village in the Swiss Alps, where he was on holidays with a young lady of his acquaintance (while his wife Anny stayed in Zurich). He started calculations at their hotel. On about the third day of the trip, Schrödinger derived the time-independent wave equation for matter waves, subsequently called the Schrödinger Equation, which became the foundation for his quantum wave mechanics. Returning to Zurich, he gave another colloquium in which he presented his new formalism of quantum mechanics to his colleagues.

He published a series of four papers laying out the new quantum formalism in detail in 1926. In the first of these [4], he presented a derivation of the time-independent Schrödinger equation, which is applicable to stationary states in which there is no change in energy. In the second paper [5], he re-derived the

**Fig. 2.9** In 1933, Erwin Schrödinger (1887–1961) received the Nobel Prize in Physics for his work on wave mechanics

time-independent Schrödinger equation and solved the problems of the rigid rotor (spinning top) and the harmonic oscillator (mass + spring vibrations). The third paper [6] addressed the equivalence of his wave mechanics and Heisenberg's matrix mechanics, and he solved the problem of an atom in an external electric field (the Stark effect). The fourth paper [7] derived the time-dependent Schrödinger equation (permitting the investigation of systems with changing energy) and addressed the problems of atomic transitions and particle scattering.

What is it that Schrödinger did to obtain his equation? A wave equation is a differential equation (an equation stating a relation between some wave function and its space and time derivatives) that has as its solutions waves that travel through space (see Appendix B.4). For light waves, the wave equation of Maxwell required that the second time-derivative of the wave function was equal to the speed of light squared times the second space-derivative of the wave function. The solutions of Maxwell's wave equation are sinusoidal electromagnetic waves that travel through space at the speed of light ($c$). Schrödinger's problem was to produce a similar wave equation, but one that would have as its solutions matter waves that had a characteristic mass $m$ and traveled through space at a slower speed $v$ appropriate to the momentum $p = mv$ and kinetic energy $E = \frac{1}{2}mv^2$ of massive particles.

He accomplished this by comparing the relationships between kinetic energy and momentum that were appropriate for the particle-waves of light and of matter. For particle-waves of light, the photon's energy equals the speed of light times the photon's momentum ($E = cp$, or more relevant to Maxwell's wave equation, $E^2 = c^2p^2$). For particle-waves of matter, kinetic energy equals the momentum squared divided by twice the mass ($E = p^2/2m$). Starting with the space and time derivatives that would extract the energy and momentum from the wave function, Schrödinger constructed a differential equation that was the equivalent of the latter relation, and this became the Schrödinger equation (see Appendix B for further details).

We can see from this account that, unlike Heisenberg, Schrödinger started from a definite picture of the underlying physics when he formulated quantum wave mechanics. His picture was that matter waves were like Maxwell's electromagnetic waves. He rejected Bohr's idea of instantaneous quantum jumps and pictured matter particles as waves moving through space from one place to another, carrying energy and momentum with them, just as do photons of light. One could visualize such matter waves moving on little trajectories through space-time, connecting one interaction with the next. Schrödinger attempted to form his waves into "wave packets" that represented particles, but he found that his packets tended to come apart as they moved.

The problem with this naïve picture is that, when carefully compared with what is known about quantum behavior, it is not consistent. In September, 1926, Bohr arranged for Schrödinger to visit Copenhagen. There Bohr, aided by Heisenberg who came to Copenhagen for the occasion, held lengthy discussions with Schrödinger, attempting to convince him that while his wave-mechanics formalism was valid, the naïve picture that he saw behind it was not [8]. Bohr argued that many aspects of quantum wave behavior, particularly the phenomena of wave function reduction, probability, and uncertainty, were not consistent with Schrödinger's simple ideas of matter waves-in-space as matter analogs of classical electromagnetic waves.

Bohr persisted in these arguments so vigorously that Schrödinger actually became physically ill, and Bohr's wife Margrethe had to nurse him back to health. However, Schrödinger was not convinced that there was any interpretational problems in his views until after a lengthy correspondence with Bohr and Heisenberg that followed the visit. He finally acquiesced, but he retained a dislike for the Copenhagen view of quantum mechanics for the rest of his life. And his new wave mechanics had to go forward without any underlying picture of what lay behind it.

The upshot of these events was that quantum mechanics, while possessing two alternative formalisms, was left without any picture of what lay behind the mathematics or of the inner mechanisms that produced quantum behavior. A pictorial interpretation of the formalism was missing, and many believed that none was possible.

The British electrical engineer and mathematical physicist Paul Dirac, on reading the papers of Schrödinger and Heisenberg describing the two new quantum mechanics formalisms, quickly recognized the importance of the rival theories and investigated their relationship. By 1927, he was able to derive a more general quantum formalism and to show that both wave mechanics and matrix mechanics could be derived from it. The problem of the two competing formalisms had been resolved, but the question of their interpretation remained.

## 2.4 Heisenberg and Uncertainty (1927)

While Heisenberg was the originator of the matrix mechanics formalism of quantum mechanics, he also carefully studied the rival wave mechanics formalism originated by Schrödinger and generalized by Dirac and Jordan. In it, he found an interesting connection that its originators had not appreciated. The wave functions in the formalism typically depend on pairs of independent variables, position and momentum or energy and time, that are multiplied together as they appear in the formalism. Niels Bohr called these "complementary variables". This pairing is a precondition for Fourier analysis, in which a function of one of these variables can be converted into a function of the other variable by summation or integration.

As a modern example of the Fourier relation between complementary variables, consider an electronics laboratory experiment in which we use a pulse generator to produce a Gaussian-shaped electrical pulse of a certain time-width $\Delta t$. We view this pulse on an oscilloscope that has fast Fourier-transform (FFT) capabilities and can display both the time and the frequency distributions of a given signal. In the time domain, (see upper Fig. 2.10) the pulse rises smoothly to its peak and then drops to near zero, showing a definite peak amplitude and time width $\Delta t$. If we use the FFT capabilities of the oscilloscope to view the distribution of frequency components of the pulse, we find that in the frequency domain (see lower Fig. 2.10) the pulse also has a Gaussian shape, rising to a peak angular frequency $\omega_0$ and then dropping to near zero, showing a maximum amplitude at $\omega_0$ and showing an angular frequency width $\Delta\omega$ around this center. Now if we change the pulse generator to increase the time width of the pulse, we will find that the frequency width decreases. Conversely, if we decrease the time width, the frequency width increases.

Because the complementary time and frequency variables are connected through a Fourier transform, the time and frequency widths of the pulse have a "see-saw" relation, with one increasing as the other decreases and *vice versa*. Further, because of the Fourier algebra, the product $\Delta t\,\Delta\omega$ of the time and frequency widths of the pulse is an invariant constant, representing a sort of "uncertainty principle" for pulses. We can have a pulse that is as sharply defined in time as we wish, but only at the expense of having a very broad distribution of frequency components. Conversely, we can have a pulse that has a very narrow band of frequency components, but only at the expense of having a very broad time width. We emphasize that a pulse with a very narrow time width is composed of a broad spectrum of frequency components and *does not have a precise frequency.*

Heisenberg discovered that when quantum waves are "localized" (see Appendix B.2) just this relationship exists between the position and momentum and between the energy and time for a particle described by the formalism of wave mechanics. He used this relationship to derive the uncertainty principle, which he published in a 1927 paper [9]. In particular, the position width (or uncertainty) $\Delta q$ of a localized particle described by wave mechanics is related to its momentum width (or uncertainty) $\Delta p$ by the uncertainty relation $\Delta p \cdot \Delta q \geq \hbar$. Similarly, its time width (or uncertainty)

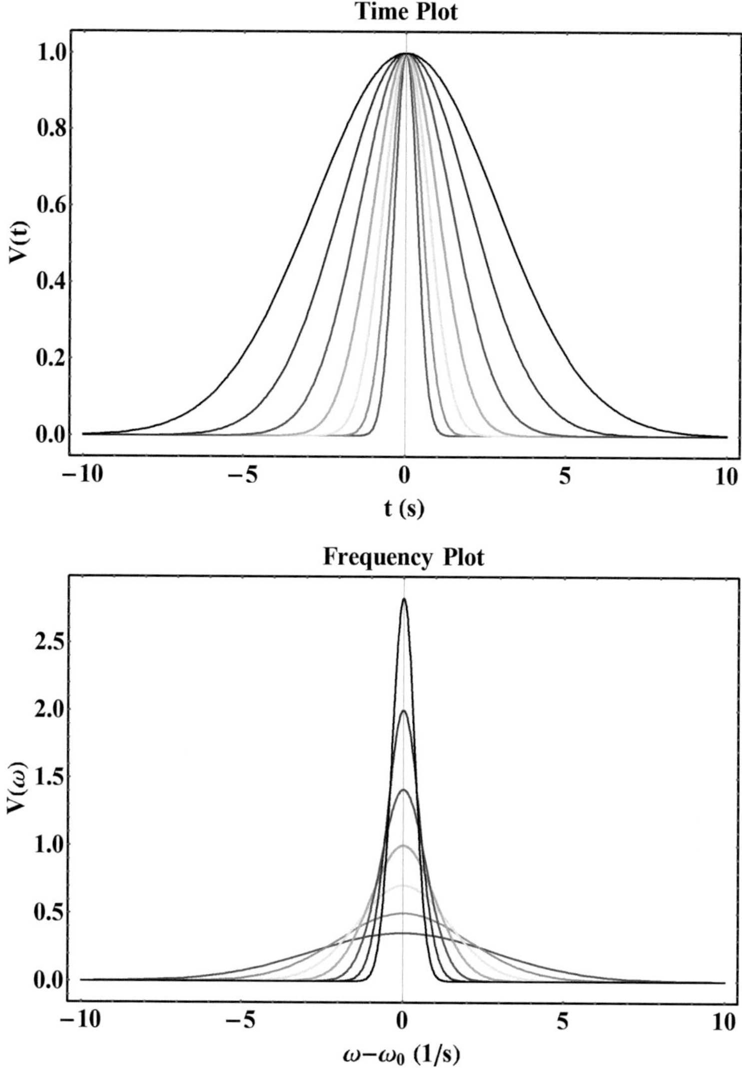


**Fig. 2.10** Gaussian voltage pulses $V(t)$ produced periodically at angular frequency $\omega_0$ with varying time widths (*narrow to broad*) and their Fourier frequency transforms $V(\omega)$ *(broad to narrow)*

$\Delta t$ is related to its energy width (or uncertainty) $\Delta E$ by the uncertainty relation $\Delta E \cdot \Delta t \geq \hbar$. This is Heisenberg's uncertainty principle.[1]

[1]More precisely, if $\sigma_x$ is the standard deviation of the position $x$ probability distribution function (PDF) and $\sigma_p$ is the standard deviation of the momentum $p_x$ PDF, then $\sigma_x \sigma_p \geq \hbar/2$. Similarly, if $\sigma_E$ is the standard deviation of the energy PDF and $\sigma_t$ is the standard deviation of the time PDF, then $\sigma_E \sigma_t \geq \hbar/2$.

Heisenberg asserted that these uncertainty relations, derived for relatively simple wave function examples, were fundamental properties of all physical systems and that they represented fundamental limits on our possible knowledge of physical quantities. We cannot simultaneously know precisely the position of a particle and its momentum (or speed or wavelength). The more precisely we can determine the value of one variable, the less precisely we can know the value of its complement. A particle with a precise position *does not have* a precise momentum value, and *vice versa*. This situation is radically different from that of classical physics, where the position, momentum, time, and energy of objects are independent variables that can be separately determined to arbitrary precision.

## 2.5 Heisenberg's Microscope (1927)

If Heisenberg had stopped there, it would have been much better for future generations of physicists. However, in his paper on the uncertainty principle [9], he chose to illustrate its operation with a *gedankenexperiment* (i.e., thought experiment), often called "Heisenberg's microscope". He envisioned that there was a particle located somewhere in empty space, and we wished to measure its position. We could do this by shooting at it a set of photons that passed through various position possibilities. When we observed that one of the photons had been Compton-scattered by the particle of interest, we would know that the particle was at the location that corresponded to that photon. The position precision of such a measurement would be restricted to wavelength of the photon. If we used high frequency radio waves, we could locate the particle with a precision of a few centimeters. If we used light, we could locate the particle with a precision of a few hundred nanometers. If we used gamma rays, we could locate the particle with a precision of a few femtometers. But each of these scattering measurements has another effect on the particle. It changes the particle's momentum by recoil, giving it a new momentum that is uncertain because we do not know all the details of its initial momentum or of the Compton scattering collision. And the shorter the wavelength of the probing photon, that larger the momentum disturbance of the particle that is struck, so the more imprecisely we can know its momentum. Position precision in the experiment is achieved only at the expense of momentum imprecision (Fig. 2.11).

Heisenberg used the increasing precision in the determination of the particle's position and the decreasing precision in our knowledge of its momentum as an illustration of the operation of the uncertainty principle. This is often called the "disturbance model", and, as first pointed out by Bohr, it is *wrong*. Nevertheless, the disturbance model is still widely used in physics textbooks by unsuspecting authors who wish to illustrate the operation of the uncertainty principle, and many generations of physics students have been led to false conclusions through its use.

The problem with the disturbance model is that it has its head wedged in classical physics. It assumes that the particle of interest simultaneously possesses a well-defined position and a well-defined momentum, and that, due to the limitations of

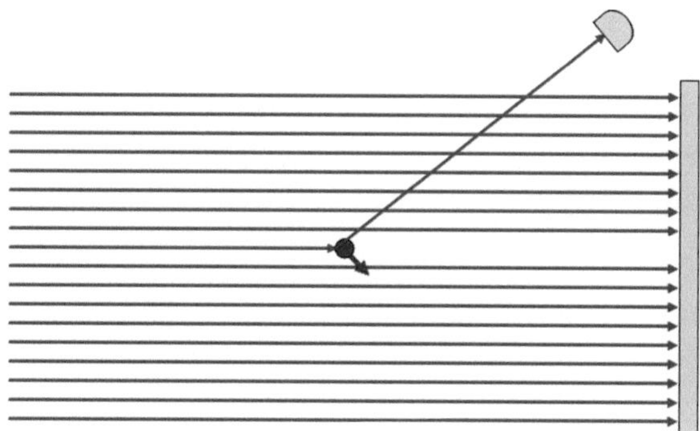

**Fig. 2.11** Heisenberg's Microscope *gedankenexperiment*: a scattered photon provides the location of a particle while disturbing potential knowledge of its momentum. Photon trajectories are one wavelength $\lambda$ apart; particle recoil is proportional to $1/\lambda$

our crude measurement capabilities, we are too clumsy to measure both of them properly at the same time.

That is not the message from the formalism of wave mechanics that Heisenberg had discovered. The complementary Fourier relations of wave mechanics tell us that a particle described by that formalism cannot possess a well-defined position and a well-defined momentum at the same time, just as our Gaussian pulse in the electronics lab illustration cannot simultaneously have a narrow time width and a narrow frequency width. They are complementary quantities, and precision in one domain makes precision in the other domain impossible. This is a fundamental property of the particle and has nothing to do with the choice or quality of measurements that we perform on it. Through measurement, we may choose to restrict the range of values that a variable may have to some arbitrary precision. The post-measurement particle, as described by the mathematical formalism, must have the precision of the complementary variable automatically expanded by the formalism to compensate, as described by Fourier algebra and the uncertainty principle. Mathematics, not measurement difficulties, is behind the uncertainty principle.

In Copenhagen, Heisenberg's paper, which had already been submitted for publication by the time Bohr read it, triggered strong and ongoing arguments at the Institute between the two theorists. Bohr argued that the aperture restricting the lateral range of the photon was of key importance, led in a simple way to the uncertainty relation, and had been ignored in favor of questionable scattering arguments. Reportedly at one point Heisenberg was reduced to tears by the strength of Bohr's arguments. As a result of the lengthy discussions at the Institute, when the page proofs for the uncertainty principle paper were received from the journal publisher, Heisenberg appended a "note added in proof", a long paragraph that rather vaguely outlined Bohr's objections to the disturbance model and admitted its inadequacies. That note, unfortunately, did not discourage later authors from lifting the disturbance model and Heisenberg's microscope *gedankenexperiment* from the publication and using them widely in the physics literature, particularly in textbooks.

## 2.6 The Copenhagen Interpretation (1927)

The quantum theoretical work of the 1924–27 period described in the previous sections had delivered a new theory of quantum mechanics that was unlike any previous physical theory. It had a well developed formalism (two, in fact), but there seemed to be no picture behind it that allowed practitioners to visualize the operation of the system they were describing with mathematics. More disturbing, the new quantum formalism had brought with it a number of unanswered questions and problems of interpretation that are still troubling physicists and philosophers, some eight and a half decades later.

Here we want to introduce these interpretational problems, not all of which were fully appreciated when Bohr, Heisenberg, and Born first developed and promoted their interpretation in late 1927. In no particular order, here is at least a partial list of such interpretational problems.

- **The problem of identity** What is the meaning of the wave function (or state vector) of wave mechanics and the matrix elements of matrix mechanics and where does it exist?
- **The problem of complexity** Why, unlike any other physical theory, are the wave function and matrix elements of quantum mechanics allowed to be complex, with both real and imaginary parts?
- **The problem of wave-particle duality** How can the mutually exclusive particle-like and wave-like behaviors described by quantum mechanical systems and observed in experiments be reconciled?
- **The problem of indeterminism** Why is the quantum formalism able to make only probabilistic, but not definite and deterministic, predictions of the outcome of well specified physical situations? How can identical conditions produce varying results? What is the source of the intrinsic randomness?
- **The problem of measurement and collapse** How and why does the wave function (or state vector) of wave mechanics change abruptly and discontinuously when a measurement is made? What is the mechanism behind state reduction?
- **The problem of nonlocality** How and why are separated but entangled parts of a quantum mechanical system nonlocally connected, so that measurements on one subsystem somehow influence the outcomes of measurements on the other subsystem, even when they are out of speed-of-light contact. What is the mechanism behind nonlocality?

As we observed in Sect. 2.2, the experience of Heisenberg with the intellectual traps implicit in pictorial models of physical systems had led him to distrust them, and to focus on the "real" variables of physical systems that could be measured in the laboratory. He carried over this approach to his interpretation of quantum mechanics. Logical positivism was fashionable in the philosophical circles of Berlin and Vienna in the late 1920s, and, perhaps influenced by his philosopher colleagues, Heisenberg adopted positivist thinking to his work [10]. He decided that one should not attempt to "look behind the scenes" at the inner workings of quantum mechanics that were inaccessible to physical measurement. These were "off limits". One should focus on observables and on the outcome of real or, if necessary, *gedanken* measurements. His uncertainty principle reinforced this view by demonstrating that many of the "virtual variables" suggested by pictorial models were not only elusive, but were completely impossible to measure, even in principle. The emerging Copenhagen Interpretation became essentially a "don't ask; don't tell" approach to the quantum formalism that fulfilled the needs of those who wanted to calculate and make predictions, but frustrated those who wanted to understand what went on behind the scenes. In later years, this emphasis on positivism was somewhat attenuated in Heisenberg's writings, but it never completely disappeared. These interpretational ideas were not explicitly called "The Copenhagen Interpretation" until 1955, when Heisenberg gave them that name [11], after which the term was widely adopted.

Niels Bohr focused on the wave-particle duality problem, emphasizing the relation between the complementarity of these aspects and the complementarity of the conjugate variables in the uncertainty principle [12]. He also emphasized the "oneness" of the system and the measurements performed on it, and insisted that these could not be separated and analyzed separately. His philosophy of complementarity, the idea that two seemingly contradictory descriptions together characterized the same phenomenon, was widely promoted and became an important part of the emerging Copenhagen Interpretation.

Max Born (Fig. 2.12), in working on the development of matrix mechanics and investigating its connection to wave mechanics, originated what became known as the "Born probability rule", the assumption that the complex values of wave functions and matrix elements could be related to physical observables by multiplying the complex quantity by its complex conjugate [13]. In other words, one added the square of the real part of the variable to the square of its imaginary part, producing a real positive number that was interpreted as the probability of making the particular observation

**Fig. 2.12** In 1954, Max Born (1882–1970) received the Nobel Prize in Physics in 1954 for his contributions to quantum mechanics

that had specified what went into the calculation. ($P = \psi\psi^*$) This became a central assumption of the Copenhagen Interpretation and an important guide for quantum mechanics practitioners who wished to relate calculations to observations. However, it led to the problem of indeterminism listed above, because, as a probability, it meant that precisely the same physical situation might have many different outcomes. The crisp deterministic character of Newtonian dynamics had been replaced by the fuzzy calculation of probabilities and by varying outcomes from identical initial conditions.

Werner Heisenberg addressed the problem of identity by asserting that the wave function was not a real wave moving through space-time, but rather was an evolving mathematical representation of *the knowledge of an observer*, real or potential, who was observing a quantum mechanical system and performing measurements on it [14, 15]. This was Heisenberg's "knowledge interpretation", which became a central element of the Copenhagen Interpretation.

At a stroke, the knowledge interpretation dealt with several of the other interpretational problems listed above: The wave function was allowed to be complex because it was an encoding of knowledge and only its absolute square, a real variable, could be directly observed. Wave-particle duality was allowed because the uncertainty principle prevented measurements that revealed particle-like and wave-like behaviors at the same time (see, however, the Afshar experiment [16] described in Sect. 6.15). The outcomes of measurements was indeterminate because of the mathematics of the formalism permitted a measurement to select a particular value from the distribution of possible values present in the wave function, that distribution representing the lack of knowledge of the observer as to the value of the measured quantity until a measurement was made. The wave function "collapsed" when the measured value became known and the knowledge of the observer changed.

The Freedman–Clauser experimental results [17], an experimental demonstration on quantum nonlocality, were published in 1972, four years before Heisenberg's death on February 1, 1976. To my knowledge, he never attempted to apply his knowledge interpretation to the emerging experimental multi-measurement demonstrations of EPR-type nonlocality, i.e. the enforced correlation of separate measurements on two separated but entangled subsystems (see Chap. 3), that implicitly would involve the knowledge of two separated observers. However, in 1984 I pressed the issue with the late Sir Rudolf Peierls (1907–1995), a Copenhagenist and skilled practitioner of Heisenberg's knowledge interpretation [18], who was visiting the University of Washington Physics Department at the time. Peierls addressed the problem of EPR nonlocality in the Freedman–Clauser experiment by observing that there was no real observer-to-observer communication involved in EPR nonlocality, only correlations. He went on to point out that the state vector *should* be allowed to be nonlocal when describing the observer's knowledge of the separated parts of the system, because the observer's knowledge must span all of the subsystems. I was looking for mechanisms, and I found this to be an interesting but unsatisfying answer.

In any case, the knowledge interpretation is a self-consistent viewpoint, even if it leaves many tantalizing quantum questions unanswered and raises many other issues (e.g., observer-created reality). The central tenets of the Copenhagen Interpretation can be summarized as follows:

- A system is completely described by a wave function $\psi$, which is a solution of a wave equation characteristic of the system; the wave function is a mathematical representation of an observer's knowledge of the system and changes when knowledge changes.
- One should focus on the observable quantities of a system and avoid asking questions about aspects that are not subject to measurement.
- The quantum mechanical description of nature is probabilistic and random. The probability of an event is the absolute square of the amplitude of the wave function related to it. Identical conditions can produce varying outcomes.
- It is not possible to know the precise values of all of the properties of a system at the same time; complementary variables are subject to the uncertainty principle; properties that are not known are described by probabilities.
- Matter and light exhibit wave-particle duality; an experiment can show the particle-like properties or wave-like properties, but *not* both at the same time. The uncertainty principle prevents wave versus particle conflicts.
- Measuring devices are essentially classical devices and measure classical properties such as position and momentum; the quantum system and the apparatus that makes measurements on it are parts of a unified whole and cannot be separated and analyzed separately.
- The quantum mechanical description of a system, in the limit of large quantum numbers, should closely correspond to its classical description.

The Fifth International Solvay Conference on Electrons and Photons, held in Brussels in October 1927, served as the coming-out party for this new way of interpreting quantum mechanics (later given the name Copenhagen Interpretation by Heisenberg [11]). Bohr, Heisenberg, and Born presented their new interpretation and defended it against all assaults. Einstein, Schrödinger, and others raised objections (see Sect. 6.2), but in the end, it was generally acknowledged that the Copenhagen Interpretation had become the standard way of approaching the quantum formalism, and that would-be practitioners of the formalism would be best guided by following its tenets.

At the Sixth International Solvay Conference on Magnetism held in Brussels in 1930, Einstein confronted the Copenhagen view by presenting his clock paradox, an ingenious *gedankenexperiment* involving a photon in a box that seemed to violate the uncertainty principle between time and energy. Leon Rosenfeld, a scientist who had participated in the Congress, described the event several years later: [19]

> It was a real shock for Bohr…who, at first, could not think of a solution. For the entire evening, he was extremely agitated, and he continued passing from one scientist to another, seeking to persuade them that it could not be the case, that it would have been the end of physics if Einstein were right; but he could not come up with any way to resolve the paradox. I will never forget the image of the two antagonists as they left the club: Einstein, with his tall and commanding figure, who walked tranquilly, with a mildly ironic smile, and Bohr who trotted along beside him, full of excitement.

The next morning brought Bohr's triumph. He presented his solution to the puzzle, an intricate account of the necessary measurements and their uncertainties that invoked Einstein's own equivalence principle to establish that the time-energy uncertainty principle was indeed preserved. Einstein conceded defeat on that occasion, but five years later in 1935 he made another assault on quantum mechanics and the Copenhagen Interpretation with the famous Einstein, Podolsky, and Rosen paper.

## 2.7 Einstein, Podolsky, and Rosen; Schrödinger and Bohm (1935–1963)

In 1935, five years after his clock paradox at the 6th Solvay Conference had failed to demonstrate the inadequacies of the uncertainty principle and Copenhagen quantum mechanics, Einstein made another foray against quantum mechanics. This time, he collaborated with colleagues Boris Podolsky and Nathan Rosen at Princeton's Institute for Advanced Studies in producing a succinct 4-page paper that described two things that he regarded as fatal flaws in quantum mechanics. They published this in the journal *Physical Review* on May 15, 1935 [20]. The work received considerable attention and soon became known as "the EPR paper". The first section of the EPR paper raised an objection to the role of the uncertainty principle in the description of quantum systems. The authors argued that if knowledge of one member of a pair of conjugate physical quantities precludes knowledge of the other quantity, then either (1) the description of reality given by the wave function in quantum mechanics is not complete, or (2) the two conjugate quantities cannot have simultaneous reality.

The second section of the EPR paper raised the issue of making quantum mechanical predictions about the states of two systems that have previously been in physical contact and are then separated. Their arguments, based on momentum conservation, focused on the choice of momentum or position measurements made on one of the systems and its effect on the quantum mechanical state of the other system. The authors argued that the choice of which quantities are measured in one system affects the outcomes of possible measurements made on the other systems. They then argued that "since the systems no longer interact, no real change can take place in the second system in consequence of anything that may be done to the first system." This seeming contradiction leads them to assert that possibility (2) cannot be true, and therefore quantum mechanics must be incomplete. The issues raised in the second section of the paper have become know as "the EPR paradox" and arise from a previously ignored aspect of the quantum mechanics formalism, its *nonlocality* or enforcement of correlations between measurements in spatially separated systems. (See Chap. 3.) In a letter to Max Born [21, 22], Einstein dismissively referred to quantum nonlocality as "spooky actions at a distance".

Niels Bohr responded to the EPR paper by defending the uncertainty principle, focusing on the simultaneous reality and indefiniteness of complementary variables like momentum and position, and discussing the Copenhagen view of wave-particle duality, while essentially ignoring the nonlocality issue that the EPR paper had raised [23]. Heisenberg's reaction to the EPR paper, as reported by his close associate C. F. von Weizsäcker [24], was: "Now, after all, Einstein has understood quantum mechanics. I am sorry for him that he still does not like it."

Schrödinger, on the other hand, took the EPR paradox and the discovery of nonlocality very seriously. He published a two-part paper in 1935-36 in which he agreed that standard quantum mechanics did indeed exhibit the property of nonlocality [25, 26]. He analyzed its aspects and implications in considerable detail. In these papers, he introduced the term *entanglement* to describe the condition of a pair of quantum systems that have interacted and then separated. He concluded that the quantum state of one of an entangled pair of systems cannot be described without making reference to the quantum state of the other member of the entangled pair. He ended by stating that "these conclusions, unavoidable within the present theory but repugnant to some physicists *including the author*, are caused by applying non-relativistic quantum mechanics beyond its legitimate range." In other words, Schrödinger took the demonstrated presence of entanglement and nonlocality in the formalism of quantum mechanics as indications that the theory must be incorrect when applied to systems where nonlocality is important.

In principle, the theory of quantum mechanics, at that point, could have been subjected to experimental testing to determine if it was indeed "being applied beyond its legitimate range". Some conserved quantity like momentum or angular momentum under different localization conditions could have been measured in two subsystems to reveal EPR correlations. In practice, in part because of the formidable experimental challenges of such tests, in part because of the lack of a crisp and falsifiable theoretical prediction, in part because World War II was in the making, and in part because of a lack of interest in such tests among experimental physicists and their sources of funding, it required almost four decades for such testing to begin.

We note, in this context, a missed opportunity: the pair of back-to-back 0.511 MeV gamma rays produced in electron-positron annihilation form a polarization-entangled photon pair because the annihilation process has zero net angular momentum and negative parity. As predicted in a calculation by John Wheeler [27], this leads to planes of linear polarization of the two photons that must be 90° apart (see Eq. 2.2).

Therefore, this system could, in principle have provided a test-bed for the testing and demonstration of EPR quantum nonlocality. C. S. Wu and I. Shaknov in 1950 [28] used a $\beta^+$-radioactive $^{64}$Cu source to measure the polarization correlations of the gamma ray pair from the $e^+e^-$ annihilation that followed the positron decay. They showed that, after efficiency and geometrical corrections, the ratio of counts with the polarimeter planes perpendicular versus parallel agreed with Wheeler's predictions based on quantum mechanics. However, since this work preceded that of J. S. Bell by a decade and a half, the measurements were only made at polarimeter angle-differences of 0°, 90°, 180°, and 270°, and no one realized the very fundamental significance of the measurements. No connection between these experimental results and Einstein's EPR arguments or nonlocality was made at the time or later in Bell's papers.

About 1951, EPR supporter David Bohm introduced the idea of a "local hidden variable" theory that could replace standard quantum mechanics with a theoretical structure that omitted the paradoxical features of quantum mechanics to which the EPR paper had objected. In Bohm's hidden-variable alternative to quantum mechanics, all correlations were established locally at sub-light speeds. Position and momentum were permitted to have simultaneous precise values, values that were real but were "hidden" and inaccessible to direct measurement.

Practicing physicists, however, paid little attention to such hidden variable theories. Bohm's approach was less useful than orthodox quantum mechanics for calculating the behavior of physical systems. Since the theories seemed to make the same predictions, it was apparently impossible to resolve the EPR/hidden-variable debate by performing an experiment, so the general physics community tended to ignore the whole controversy and leave it to the philosophers.

## 2.8 Bell's Theorem and Experimental EPR Tests (1964–1998)

In 1964, the testability situation changed. In a series of publications, John Stuart Bell (Fig. 2.13), a Scottish theoretical particle physicist working at the CERN high-energy physics laboratory in Geneva, proved an amazing theorem demonstrating that experimental tests could distinguish the predictions of quantum mechanics from those of any local hidden-variable theory [29, 30]. Bell, following the lead of Bohm [31], had based his calculations not on measurements of position and momentum, the focus of the arguments of Einstein and Schrödinger, but on measurements of the linear polarization of photons of light when considerations of angular momentum conservation constrained them.

Before discussing Bell's theorem further, we pause to consider the polarization of light. The phenomena we refer to as light: visible light, invisible infrared or ultraviolet rays, radio waves, X-rays, or gamma rays, all are aspects of the same basic physics, differing only in frequency. They are traveling waves produced when electric and magnetic fields vibrate together at right angles to each other as they move through space at the speed of light. The direction in which the electric field of a light wave vibrates determines the *polarization* of the wave. If the electric field vibrates always in the same plane, we say that this is the plane of polarization and that the wave has *linear polarization* in that plane. There is another polarization basis, *circular polarization*, in which the electric field corkscrews through space to the right or left as the light wave moves forward. Circular polarization can be produced by superimposing states of vertical and horizontal linear polarization with appropriate phase, and linear polarization can be produced by superimposing left and right circular polarization. (See Eqs. 6.2–6.5.) Most of the work with Bell's theorem has been focused on states of linear polarization.

**Fig. 2.13** John Stuart Bell (1928–1990)

It is quite easy to measure the linear polarization of visible light. Special optical filters, for example the lenses of polarized sunglasses, absorb light polarized in one direction while transmitting the light polarized in the perpendicular direction. There are also polarization-sensitive splitters that divide one beam of light into two beams, for example with one linear polarization state reflected and the other transmitted.

Using such devices, a particular polarization component of incident light can be transmitted and the other component absorbed or diverted. If a beam of unpolarized light is passed first through one such polarization filter and then through another, the intensity of the transmitted beam varies in accordance with Malus' Law, which states that the transmitted light intensity $I(\alpha)$ is proportional to the square of the cosine of the angle $\alpha$ between the polarization direction of the first filter and that of the second filter, i.e., $I(\alpha) = I_0 \cos^2(\alpha)$, where $I_0$ is the intensity observed when the polarization directions of the filters are parallel. Malus' Law is shown in Fig. 2.14 This equation tells us that when the planes of polarization of the two filters are at 90°, the crossed filters look black and no light is transmitted. When the planes of polarization make an angle of 45°, half the light intensity passing the first filter is transmitted by the second, and so on.

In an atom, if an orbital electron is kicked from its lowest energy level into a higher orbit by an energetic photon or an electrical discharge, the electron may return to its lowest energy state by a process called a "cascade", a series of quantum jumps to lower orbits, each jump producing a single light photon of a wavelength that depends on the energy gap of the jump. A two-photon cascade in which the atom as a whole begins and ends with no net angular momentum (i.e., no rotational motion) and no change in parity (the mirror-symmetry of the system is unchanged) is of particular interest, because the cascade produces an entangled pair of photons

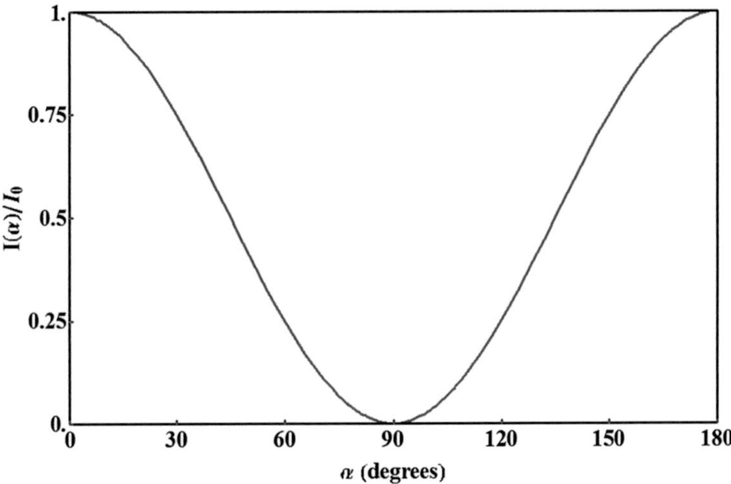


**Fig. 2.14** Malus' Law: Linearly polarized light is transmitted through two polarizing filters with intensity $I(\alpha) = I_0 \cos^2(\alpha)$, where $\alpha$ is the angle between the polarizing axes of the filters

that have correlated polarizations due to angular momentum conservation. When the photons from the cascade travel back-to-back in opposite directions, angular momentum conservation requires that if one of the photons is measured to have some definite linear polarization state, the other photon is required to have exactly the same linear polarization state. Kocher and Commins [32] used this technique to produce entangled photon pairs and to demonstrate the polarization correlation predicted by quantum mechanics, but they did not attempt a test of Bell's theorem. The first experimental tests of Bell's theorem in the 1970s, often called "EPR experiments", used the entangled photon pairs from such cascades.

First, a word about notation: in the discussions that follow we will explicitly indicate wave functions $\psi$ using the Dirac "ket" state vector notation; a *ket* is a vertical bar $|$ and an angle bracket $\rangle$ that enclose some symbol that distinguishes one wave function from another. For example, a wave function that is characteristic of system 1 in a state of horizontal linear polarization might be represented by $\psi_{H_1} = | H\rangle_1$, and so on. Later we will also use the complex conjugate of the Dirac ket state vector, which is called a *bra* and for the ket above would be denoted by the symbol $\langle H |_1$.

The wave function that describes such an entangled pair of photons is said to describe a "Bell state", and, for zero initial and final angular momentum, if there is no change in parity it has the symmetric form:

$$| S\rangle_+ = \frac{1}{\sqrt{2}}(| H\rangle_1 | H\rangle_2 + e^{i\phi} | V\rangle_1 | V\rangle_2), \tag{2.1}$$

or if parity changes, it has the anti-symmetric form:

$$\mid S\rangle_{-} = \frac{1}{\sqrt{2}}(\mid H\rangle_1 \mid V\rangle_2 - e^{i\phi} \mid V\rangle_1 \mid H\rangle_2), \tag{2.2}$$

where $\phi$ is an arbitrary phase angle that depends on geometry and is usually 0 or $\pi$, and $H$ and $V$ describe horizontal and vertical linear polarization, respectively, of the entangled photons moving on paths 1 and 2. When a Bell-state wave function described by Eq. 2.1 is collapsed by a linear polarization measurement of the photons, it will be found that either both photons have $H$ polarization or that both photons have $V$ polarization, each with a 50 % probability.

These EPR experiments measured the coincident arrival of entangled photons at opposite ends of the apparatus, as detected by quantum-sensitive photomultiplier tubes after each photon had passed through a polarizing filter. The photomultipliers at opposite ends of the apparatus produce electrical pulses that, when they occur at the same time, are recorded as a "coincidence" or two-photon event. The rate of such coincident events is measured while varying the polarization "pass" directions of the two filters, characterized by transmission-axis angles $\alpha_1$ and $\alpha_2$. The two transmission angles are systematically varied and the rate measurement is repeated until a complete map of rate versus the two angles is developed.

Bell's theorem deals with the way in which the coincidence rate of an EPR experiment falls off when the two transmission angles $\alpha_1$ and $\alpha_2$ are not equal. Bell proved mathematically that for all local hidden-variable theories [31] the magnitude of the decrease in coincidence rate must be linear (or less) as it depends on the angular difference $\Delta\alpha$ between the two filters. Suppose, for example, that we misalign the angles of the two polarization filters so that the angle between the polarization directions of the two filters is $\Delta\alpha = \alpha_1 - \alpha_2$. We measure the coincidence rate $R(\Delta\alpha)$, as compared to the rate $R_0$ when the filters are perfectly aligned. That rate drops by an amount $\Delta_1 = R_0 - R(\Delta\alpha)$. Now we double the amount of the misalignment, so that the decrease in rate is $\Delta_2 = R_0 - R(2\Delta\alpha)$. For this situation, Bell's theorem requires that $\Delta_2$ must be less than or equal to twice $\Delta_1$ ($\Delta_2 \leq 2\Delta_1$).

This prediction of Bell's theorem is one of the so-called "Bell inequalities". It can be thought of in the following way. Consider that the coincidence rate $R_0$ when the polarizing filters are aligned ($\Delta\alpha = 0$) is a "signal", to which "noise" is added when a misalignment is introduced. If the noise $\Delta_1$ introduced by moving one filter an amount $\alpha$ to the right is not correlated with the noise $\Delta_1'$ introduced by moving the other filter by the same angle to the left, then at most, when both sources of noise are present, the noise $\Delta_2$ from a $2\alpha$ misalignment should be twice $\Delta_1$. However, the two uncorrelated noise sources may occasionally cancel, permitting $\Delta_2$ to be less than twice $\Delta_1$. Therefore, the Bell inequality states that $\Delta_2$ must be less than or equal to twice $\Delta_1$.

Quantum mechanics, on the other hand, predicts that the coincidence rate $R(\alpha_1, \alpha_2)$ depends only on the relative angle $\Delta\alpha = \alpha_1 - \alpha_2$ between the two polarization directions, and that $R(\Delta\alpha)$ obeys Malus' Law. In other words, quantum mechanics predicts that $R(\alpha_1, \alpha_2) = R(\Delta\alpha) = R_0 \cos^2(\Delta\alpha)$. Therefore, $\Delta_1 = R_0[1 - \cos^2(\Delta\alpha)]$ and $\Delta_2 = R_0[1 - \cos^2(2\Delta\alpha)]$. When the misalignment angle $\alpha$ is fairly small, this means that $\Delta_2$ is about four times $\Delta_1$, which is clearly much larger than twice $\Delta_1$ (i.e., $\Delta_2 \approx 4\Delta_1$ so $\Delta_2 > 2\Delta_1$). This is a clear violation of Bell's theorem, because the coincidence rate predicted by quantum mechanics falls off much too fast with increasing angle to be consistent with Bell's theorem, which predicts an approximately linear decrease, as shown in Fig. 2.15.

What is the essential difference between quantum mechanics and local hidden-variable theories that causes their distinguishable predictions of the relation between $\Delta_1$ and $\Delta_2$? As pointed out by Herbert [33], in the local hidden-variable theories, the photons leaving the source are required to be in a definite (but possibly random and unknown) state of linear polarization, leading to noise $\Delta$ roughly proportional to $\Delta\alpha$. For quantum mechanics, the state of polarization of entangled photons leaving the source is indefinite and is not fixed until a polarization measurement is made, leading to Malus' Law and $\Delta$ roughly proportional to $\Delta\alpha^2$. (See the discussion of "realism" in Sect. 6.18). That essential difference between linear and quadratic behavior lies at the root of Bell's inequalities.

We note that Bell did not consider the less-than-100 % efficiency of real single photon detectors and the less-than-perfect behavior of real polarization analyzers. A group led by John Clauser [34] generalized Bell's theorem to take these effects into account, producing the CHSH inequality, which is used in real experimental tests and is essentially Bell's inequality cast in a more realistic experimental context.

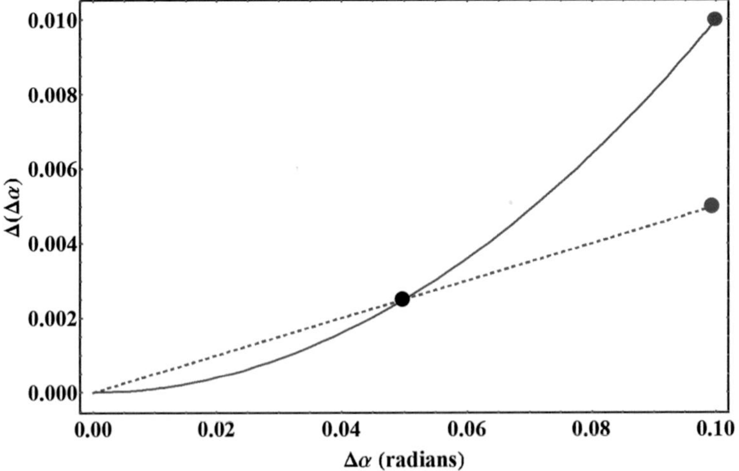


**Fig. 2.15** Plot of "noise" rate $\Delta(\Delta\alpha)$ versus polarimeter angle difference $\Delta\alpha$ for small angles. Local hidden variable prediction (*blue/dashed*) rises linearly, while quantum mechanical prediction (*red/solid*) rises quadratically. If the rates are equal at $\Delta\alpha = 0.05$ (*black dot*), the quantum mechanical prediction (*red dot*) is twice the local hidden variable prediction (*blue dot*) at $\Delta\alpha = 0.10$ and four times the rates at $\Delta\alpha = 0.05$

The first unambiguous experimental results from EPR experiments, the pioneering work of John Clauser (Fig. 2.16) and his student Stuart Freedman (Fig. 2.17) at UC Berkeley, was performed in the early 1970s and published in 1972 [17]. They reported a 6.7 standard deviation violation of Bells' inequalities and consistency with the predictions of quantum mechanics. A decade later, in 1982, the EPR measurements of the Aspect group [35, 36] in France used newly developed apparatus and techniques and were able to eliminate several "loophole" scenarios that might constitute unlikely ways of preserving classical locality (see Sect. 3.2). They demonstrated consistency with quantum mechanics and inconsistency with local hidden-variable theories, this time with a 46 standard deviation violation of Bell's inequalities [35, 36]. A more recent EPR experimental example is the 1998 work of the Gisin group in Switzerland [37, 38]. They used fiber-optics cables owned by the Swiss Telephone System to demonstrate the nonlocal connection between EPR measurements made at locations in Geneva and Bern, Swiss cities with a line-of-sight separation of 156 km. Their work constitutes a direct demonstration, if one was required, that not only is quantum mechanics nonlocal, but that such nonlocality can operate over quite large spatial separations.

Such EPR results were initially interpreted as a demonstration that hidden variable theories like those of Bohm had been falsified. That view changed when it was realized that Bell's theorem was based on *local* hidden variable theories, and that nonlocal hidden variable theories could also be constructed to violate Bell's theorem and agree with the experimental measurements. The assumption made by Bell that had been put to the test was the assumption of locality, not hidden variables.

Do these EPR experiments constitute a solid demonstration of the existence of quantum nonlocality? There is more than one way of interpreting the implications of the experimental results, and one can find much discussion in the literature as

**Fig. 2.16** John F. Clauser (1942–)

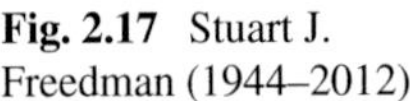

**Fig. 2.17** Stuart J. Freedman (1944–2012)

to whether it is locality or "realism" (the objective observer-independent reality of external events) that has been refuted by these EPR measurements. We here adopt the view that reality should be taken as a given, and so we regard these experiments as direct demonstrations of the intrinsic nonlocality of standard quantum mechanics. (See Sect. 6.18 for further discussion of the realism issue.)

The EPR experiments demonstrate the quantum enforcement of a correlation between measurement results in the two separated arms of the experiment. Let us try to clarify the nature of that correlation with an example. Suppose that you are given a gold coin, and you use a fine jeweler's saw to cut it in half along the plane of the coin, placing the "head" side in one pink envelope and the "tail" side in another. You do the same thing with another coin, separating it with a horizontal cut into a top half and a bottom half and placing these in green envelopes. And you cut a third coin, separating it with a vertical cut into a left half and a right half and placing these in blue envelopes. Now you shuffle the envelopes and send one set of colors to an observer in Boston and the other set to an observer in Seattle. Each observer is allowed to choose one of the envelopes and to open only that one. If the Seattle observer opens his pink envelope and finds the head, he knows that the Boston observer, if he also opens his pink envelope, will find a tail. But he is unable to predict what the Boston observer will find if he opens the green or the blue envelope.

That is classical physics. The difference in the EPR quantum situation is that there is only one coin, and that each observer decides the direction in which his half of the single coin should have been be cut only after the single white envelope has arrived. And yet, he observes the same correlations as described above. If both observers choose the same cut directions, their halves are opposites. If they choose different cut directions, they are unable to predict the observation of the other observer. Here the correlated coin-halves correspond to entangled photons and the cut directions to the choice of measurements of polarization bases (circular right/left, linear vertical/horizontal, linear 45° diagonal/anti-diagonal, and others). See the quantum games described in Appendix C for further analogies to EPR correlations.

We note that the several polarization bases used in these kinds of polarization EPR experiments make it straightforward to demonstrate the quantum nonlocal connections but also make it effectively impossible to use those connections for observer-to-observer signaling, because one would need to deduce from the arriving photons the polarization basis that was being used in the distant measurements. While each observer is free to choose the polarization basis (e.g., circular right/left, linear horizontal/vertical, linear 45° diagonal/anti-diagonal) for the measurement, he is not free to force the photon into a particular state of that basis, as would be required for nonlocal communication. Thus, while polarization-based EPR experiments may be taken as demonstrations that Nature is using some nonlocal mechanism to arrange the correlations of the separated measurements, such a "superluminal telegraph line" is not accessible to the experimenters for sending their own messages. See Chap. 7 for further discussion of the suppression of nonlocal signaling.

To put it another way, the intrinsic nonlocality of quantum mechanics had been tested by the experimental EPR tests of Bell's theorem. It has been experimentally demonstrated that Nature arranges the correlations between the polarizations of the two entangled photons at separated measurement sites by some nonlocal (and perhaps retrocausal) mechanism that violates Einstein's intuitions about the intrinsic locality of all natural processes. What Einstein called "spooky actions at a distance" are in fact an important part of the way Nature works at the quantum level.

## References


1. M. Born, W. Heisenberg, Über quantentheoretische Umdeutung kinematischer und mechanischer Beziehungen. Z. für Phys. **33**, 879–893 (1925). First paper on the matrix mechanics formulation of quantum mechanics
2. M. Born, P. Jordan, Zur Quantenmechanik, Z. für Phys. **34**, 858–888 (1925). Second paper on the matrix mechanics formulation of quantum mechanics
3. M. Born, W. Heisenberg, P. Jordan, Zur Quantenmechanik II. Z. für Phys. **35**(8–9), 557–615 (1926). Third paper on the matrix mechanics formulation of quantum mechanics
4. E. Schrödinger, Ann. der Phys. **79**, 361 (1926)
5. E. Schrödinger, Ann. der Phys. **79**, 486 (1926)
6. E. Schrödinger, Ann. der Phys. **79**, 734 (1926)
7. E. Schrödinger, Ann. der Phys. **81**, 109 (1926)
8. W. Heisenberg, Reminiscences from 1926 and 1927, in reference [39]
9. W. Heisenberg, Z. für Phys. **43**, 172 (1927). Translated in [40], pp. 62–84
10. W. Heisenberg, Z. für Phys. **33**, 879 (1925)
11. W. Heisenberg, The development of the interpretation of quantum theory, in *Niels Bohr and the Development of Physics*, ed. by W. Pauli (Pergamon, London, 1955)
12. N. Bohr, *Atti del Congresso Internazionale dei Fisici Como*, 11–20 Settembre 1927, vol. 2 (Zanchelli, Bologna, 1928), pp. 565–588
13. M. Born, Z. für Phys. **37**, 863 (1926), pp. 52–55. Translated in [40]
14. W. Heisenberg, Daedalus **87**, 95 (1958)
15. M. Jammer, *The Philosophy of Quantum Mechanics* (Wiley, New York, 1974)

16. S.S. Afshar, Violation of the principle of complementarity, and its implications, Proc. SPIE, **5866**, 229–244 (2005). arXiv:quant-ph/0701027
17. S.J. Freedman, J.F. Clauser, Phys. Rev. Lett. **28**, 938 (1972)
18. R. Peierls, In defense of measurement. Phys. World 19–20 (1979)
19. L. Rosenfeld, in [40] (1955), pp. 477–478
20. A. Einstein, B. Podolsky, N. Rosen, Phys. Rev. **47**, 777–785 (1935)
21. A. Einstein in a 1947 letter to M. Born (see [22])
22. M. Born, A. Einstein, in *The Born-Einstein Letters*, ed. by M. Walker (Born, New York, 1979)
23. N. Bohr, Phys. Rev. **48**, 696 (1935)
24. Th Görnitz, C.F. von Weizsäcker, Copenhagen and transactional interpretations. Int. J. Theor. Phys. **27**, 237–250 (1988)
25. E. Schrödinger, Proc. Camb. Philos. Soc. **31**, 555–563 (1935)
26. E. Schrödinger, Proc. Camb. Philos. Soc. **32**, 446–451 (1936)
27. J.A. Wheeler, Polyelectrons. Ann. N. Y. Acad. Sci. **48**, 157 (1945)
28. C.S. Wu, I. Shaknov, The angular correlation of scattered annihilation radiation (letter to the editor). Phys. Rev. **77**, 136 (1950)
29. J.S. Bell, Physics **1**, 195 (1964)
30. J.S. Bell, Rev. Mod. Phys. **38**, 447 (1966)
31. D. Bohm, *Causality and Chance in Modern Physics*, (Routledge & Kegan Paul and D. van Nostrand, 1957). ISBN: 0-8122-1002-6
32. C.A. Kocher, E.D. Commins, Phys. Rev. Lett. **18**, 575 (1967)
33. N. Herbert, Am. J. Phys. **43**, 315 (1975)
34. J.F. Clauser, M. Horne, A. Shimony, R. Holt, Proposed experiment to test local hidden-variable theories. Phys. Rev. Lett. **23**, 880 (1969)
35. A. Aspect, J. Dalibard, G. Roger, Phys. Rev. Lett. **49**, 91 (1982)
36. A. Aspect, J. Dalibard, G. Roger, Phys. Rev. Lett. **49**, 1804 (1982)
37. W. Tittel, J. Brendel, B. Gisin, T. Herzog, H. Zbinden, N. Gisin, Experimental demonstration of quantum-correlations over more than 10 kilometers, Phys. Rev. A **57**, 3229 (1998), arXiv:quant-ph/9707042
38. W. Tittel, J. Brendel, H. Zbinden, N. Gisin, Violation of Bell inequalities by photons more than 10 km apart, Phys. Rev. Lett. **81**, 3563–6 (1998), arXiv:quant-ph/9806043
39. A.P. French, P.J. Kennedy (eds.), *Niels Bohr, A Centernary Volume* (Harvard UniversityPress, Cambridge, 1985)
40. J.A. Wheeler, W.H. Zurek (eds.) *Quantum Theory and Measurement* (Princeton University-Press, Princeton, 1983)

# Chapter 3
# Quantum Entanglement and Nonlocality

Quantum mechanics, our standard theoretical model of the physical world at the smallest scales of energy and size, differs from the classical mechanics of Newton that preceded it in one very important way. Newtonian systems are always *local*. If a Newtonian system breaks up, each of its parts has a definite and well-defined energy, momentum, and angular momentum, parceled out at breakup by the system while respecting conservation laws. After the component subsystems are separated, the properties of any subsystem are completely independent and do not depend on those of the other subsystems.

On the other hand, quantum mechanics is *nonlocal*, meaning that the component parts of a quantum system may continue to influence each other, even when they are well separated in space and out of speed-of-light contact. As discussed in Sect. 2.7, this characteristic of standard quantum theory was first pointed out by Albert Einstein and his colleagues Boris Podolsky and Nathan Rosen (EPR) in 1935, in a critical paper [1] in which they held up the discovered nonlocality as a devastating flaw that, it was claimed, demonstrated that the standard quantum formalism must be incomplete or wrong. Einstein called nonlocality "spooky actions at a distance" [2]. Schrödinger followed on the discovery of quantum nonlocality by showing in detail how the components of a multi-part quantum system must depend on each other, even when they are well separated [3, 4], but expressed doubts about the validity of nonlocality.

Beginning in 1972 with the pioneering experimental work of Stuart Freedman and John Clauser [5], a series of quantum-optics EPR experiments testing Bell-inequality violations [6] and other aspects of entangled quantum systems have been performed. This body of experimental results can be taken as a demonstration that, like it or not, both quantum mechanics and the underlying reality it describes are intrinsically nonlocal. Einstein's spooky actions-at-a-distance are really out there in the physical world, whether we understand and accept them or not.



J.G. Cramer, *The Quantum Handshake*, DOI 10.1007/978-3-319-24642-0_3

## 3.1 Conservation Laws at the Quantum Scale

As quantum mechanics was being developed in the 1920s, there were serious questions about whether energy and other quantities were actually conserved at the quantum scale. After the French physicist Henri Becquerel accidentally discovered radioactivity in 1896 while investigating the phosphorescence of uranium salts, further studies had shown that there were three distinctly different types of energetic emissions produced in radioactive decay processes: alpha-particles ($\alpha$), beta-particles ($\beta$), and gamma-rays ($\gamma$). Ongoing research with cloud chambers and Geiger counters indicated that $\alpha$-particles had a positive charge and a mass on the scale of atoms, $\gamma$-rays were electrically neutral and behaved like high energy x-rays, and $\beta$-particles had an electric charge and about the mass of electrons. Indeed, $\beta^-$ particles were later shown to be electrons and $\alpha$-particles to be helium nuclei.

But $\beta$-particles presented another problem. A radioactive nucleus of a given mass and spin would emit a $\beta$-particle, and the resulting system, consisting of a residual nucleus and a $\beta^-$ would break the laws of energy and angular momentum conservation. In particular, instead of the $\beta^-$ being emitted with a fixed energy that represented the mass-energy difference between the masses of the initial nucleus and the final nucleus, the particles were emitted with a range of energies forming a "bump" energy-distribution that started near zero and ended at the nuclear mass-energy difference. Energy was missing. Further, nuclei starting with boson-type integer spin statistics or fermion-type half-integer spin statistics before the decay did not change to the opposite spin-statistics following the decay, even though the emitted $\beta$ particle was carrying off $\hbar/2$ of angular momentum. It appeared that neither energy nor angular momentum was being conserved at the quantum scale in $\beta$-decay processes.

Niels Bohr was ready to abandon energy conservation and the idea of photons, and he turned to investigating models in which energy and angular momentum were only conserved as a statistical average over many events [7]. Bohr's Dutch assistant, Hans Kramers (Fig. 3.1), had done a calculation at Bohr's Institute in which he took Einstein's photon ideas seriously and used energy and momentum conservation to show that photons should scatter from electrons and lose energy, emerging from the scattering with a longer wavelength. Bohr, wary of the very idea of photons and convinced that energy conservation was inappropriate at the scale of individual atoms, argued strongly against Kramers' model and, after much argument, convinced him that his predictions were nonsense and should not be published. A few years later in 1923, the American experimentalist Arthur H. Compton (Fig. 3.2) of Washington University in St. Louis created a sensation in the physics community with his discovery of the Compton effect. The Compton effect was just the phenomenon that Kramers' calculations had predicted years earlier. Compton won the Nobel Prize in Physics in 1927 for his discovery, but Kramers received no recognition for having anticipated the work.

The second part of the EPR paper (see Sect. 2.7) used the conservation of momentum to argue that quantum mechanics appeared to be nonlocal. In Bohr's published response to the EPR paper [8], he completely ignored the second part of the paper and

**Fig. 3.1** Hendrik A. "Hans" Kramers (1894–1952) was Bohr's assistant for many years before becoming a Professor at the University of Leiden

focused entirely on the first part, which dealt with issued of wave-particle duality. It seems likely that Bohr took this approach because, even as late as 1935, he still distrusted momentum conservation at the microscopic scale and rejected ideas based on it.

The apparent violation of conservation laws in $\beta$-decay was subsequently explained by Wolfgang Pauli and Enrico Fermi. In 1930, Pauli, over Bohr's objections, hypothesized the existence of an unseen particle (later named the *neutrino*) that had a mass much smaller than that of an electron, was electrically neutral, and had half-integer spin. Pauli argued that a three-body $\beta$-decay into a neutrino, an electron, and a residual nucleus could permit the preservation of energy, momentum, and angular momentum conservation. Subsequently, in 1934 Fermi [9] used phase-space arguments to show that the decay of a radioactive nucleus into a three-particle final state (nucleus + electron + neutrino) could account for the observed energy distributions of $\beta$ particles. Neutrinos from a nuclear reactor were experimentally observed in 1956 [10] using inverse electron capture, confirming these ideas. Thus, despite Bohr's doubts, energy, momentum, and angular momentum are conserved in microscopic processes on an event-by-event basis, just as they are in the macroscopic world.

**Fig. 3.2** Arthur H. Compton (1892–1962) received the Nobel Prize in Physics in 1927 for his discovery of the Compton effect

## 3.2 Hidden Variables and EPR Loopholes

David Bohm invented local hidden variable theories [11, 12], not because there was a need for an alternative to standard quantum mechanics, but because, as an advocate of Einstein's point of view, he distrusted the messages of indeterminism, uncertain variables, and nonlocal behavior that were aspects of the standard quantum formalism, as viewed through the lens of the Copenhagen Interpretation. His approach was deliberately contrived to duplicate the predictions of experimental outcomes by quantum mechanics, but, even in the view of Bohm himself, it was more of a demonstration that other more palatable theories were possible than it was intended as a serious alternative to the standard quantum theory. When Bell demonstrated that local hidden variables could be falsified by experiment and EPR experiments demonstrated problems with locality, Bohm responded by devising the Bohm-de Broglie hidden-variable formalism [13] that was explicitly nonlocal (but unfortunately nonlocal in a way that conflicted with special relativity), and that tracked quantum mechanics in violating Bell's inequalities. In this way, it was demonstrated that the EPR tests had falsified locality, not hidden variables.

Following the experimental work of Freedman and Clauser [5], a curious sociological phenomenon occurred among quantum theorist and philosophers. Rather than accepting the EPR experimental work as evidence that nonlocality had to be taken seriously, a number of theorists expended much effort in cataloging EPR "loopholes": implausible scenarios involving equipment inefficiencies, sampling irregularities, hypothetical slower-than-light connections, and other classical mechanisms by which the EPR experimental results might occur without the presence of quantum nonlocality and entanglement. This created an "industry" among quantum optics experimentalists, who have spent several decades designing and performing experiments that eliminate one hypothetical loophole after another.

In the view of the author, these are experimental efforts that would be better devoted to "pushing the quantum envelope" by exploring how nonlocality and entanglement work and how they can be used. In Chap. 6 we will consider many experiments in the latter category, and we will largely ignore the loophole chasers.

## 3.3 Why Is Quantum Mechanics Nonlocal?

How and why is quantum mechanics nonlocal? Nonlocality comes from two seemingly conflicting aspects of the quantum formalism: (1) energy, momentum, and angular momentum, important properties of light and matter, are conserved in all quantum systems, in the sense that, in the absence of external forces and torques, their net values must remain unchanged as the system evolves, while (2) in the wave functions describing quantum systems, as required by Heisenberg's uncertainty principle [14], the complementary conserved quantities may be indefinite and unspecified and typically can span a large range of possible values. This non-specificity persists until a measurement is made that "collapses" the wave function and fixes the measured quantities with specific values. These seemingly inconsistent requirements of (1) and (2) raise an important question: how can the wave functions describing the separated members of a system of particles, which may be light-years apart, have arbitrary and unspecified values for the conserved quantities and yet respect the conservation laws when the wave functions are collapsed?

This paradox is accommodated in the formalism of quantum mechanics because the quantum wave functions of particles are *entangled*, the term coined by Schrödinger [3] to mean that even when the wave functions describe system parts that are spatially separated and out of light-speed contact, the separate wave functions continue to depend on each other and cannot be separately specified. In particular, the conserved quantities in the system's parts (even though individually indefinite) must always add up to the values possessed by the overall quantum system before it separated into parts. We should emphasize, however, that entangled subsystems do not *need* to be widely separated and out of light-speed contact. Entangled subsystems can be close together or far apart.

How could this entanglement and preservation of conservation laws possibly be arranged by Nature? The mathematics of quantum mechanics gives us no answers to this question, it only insists that the wave functions of separated parts of a quantum system do depend on each other. Theorists prone to abstraction have found it convenient to abandon the three-dimensional universe and describe such quantum systems as residing in a many-dimensional Hilbert hyper-space in which the conserved variables form extra dimensions and in which the interconnections between particle wave functions are represented as allowed sub-regions of the overall hyper-space. That has led to elegant mathematics, but it provides little assistance in visualizing what is really going on in the physical world. However, as we will see in Chap. 5 the Transactional Interpretation of quantum mechanics easily describes the underlying mechanism that produces nonlocality in quantum processes.

## 3.4 Nonlocality Questions Without Copenhagen Answers

Consider these questions, which are raised by quantum nonlocality (see Appendix A.2):

- Can the quantum wave functions of entangled systems be objects that exist in normal three-dimensional space?
- What are the true roles of the observers and measurements in quantum processes that involve several separated measurements on entangled subsystems?
- What is wave function collapse (or state vector reduction) and how does it occur, particularly for entangled systems?
- How can quantum nonlocality be understood?
- How can quantum nonlocality be visualized?
- What are the underlying physical processes that make quantum nonlocality possible?

To our knowledge, the only interpretation that adequately answers all of these questions about nonlocality is the Transactional Interpretation of quantum mechanics [15–19], which will be described in Chap. 5. These questions are answered from the Transactional Interpretation point of view in Appendix A.2.

## References

1. A. Einstein, B. Podolsky, N. Rosen, Phys. Rev. **47**, 777–785 (1935)
2. M. Born, A. Einstein, *The Born-Einstein Letters, with Comments by M. Born* (Walker, New York, 1979)
3. E. Schrödinger, Proc. Camb. Philos. Soc. **31**, 555–563 (1935)
4. E. Schrödinger, Proc. Camb. Philos. Soc. **32**, 446–451 (1936)
5. S.J. Freedman, J.F. Clauser, Phys. Rev. Lett. **28**, 938 (1972)
6. J.S. Bell, Physics **1**, 195 (1964)

7. A.P. French, P.J. Kennedy (eds.), *Niels Bohr, A Centernary Volume* (Harvard University Press, Cambridge, 1985)
8. N. Bohr, Phys. Rev. **48**, 696 (1935)
9. E. Fermi, Versuch einer Theorie der $\beta$-Strahlen. I. Z. Phys. A **88**, 161 (1934)
10. C.L. Cowan Jr., F. Reines, F.B. Harrison, H.W. Kruse et al., Detection of the free neutrino: a confirmation. Science **124**(3212), 1034 (1956)
11. D. Bohm, A suggested interpretation of the quantum theory in terms of hidden variables I. Phys. Rev. **85**, 166179 (1952)
12. D. Bohm, *Causality and Chance in Modern Physics* (Routledge & Kegan Paul and D. Van Nostrand, London, 1957). ISBN: 0-8122-1002-6
13. D. Bohm, B.J. Hiley, *The Undivided Universe—An Ontological Interpretation of Quantum Theory* (Routledge, London, 1993). ISBN: 0-415-06588-7
14. W. Heisenberg, Z. für Phys. **43**, 172 (1927) (translated in [20], pp. 62–84)
15. J.G. Cramer, The transactional interpretation of quantum mechanics. Rev. Mod. Phys. **58**, 647–687 (1986)
16. J.G. Cramer, An overview of the transactional interpretation of quantum mechanics. Int. J. Theor. Phys. **27**, 227–236 (1988)
17. J.G. Cramer, The plane of the present and the new transactional paradigm of time. Chapter 9, in *Time and the Instant*, ed. by R. Drurie (Clinamen Press, Manchester, 2001), arXiv:quant-ph/0507089
18. J.G. Cramer, Found. Phys. Lett. **19**, 63–73 (2006)
19. J.G. Cramer, *The Transactional Interpretation of Quantum Mechanics and Quantum Nonlocality* (to be published as a book chapter), arXiv:1503.00039 [quant-ph]
20. J.A. Wheeler, W.H. Zurek (eds.), *Quantum Theory and Measurement* (Princeton University Press, Princeton, 1983)

# Chapter 4
# Reversing Time

## 4.1 The History of $i$

When you square any number by multiplying it by itself, whether it is positive or negative, the result is always a positive number. Therefore, the square root of minus one, or $i$ in the usual mathematical notation, in some sense is meaningless or imaginary, because no real number has that mathematical property. But $i$ turns out to be useful in mathematics, physics, and engineering as a mathematical object. Nevertheless, it required some time for $i$ to assume its proper place in mathematics.

In 1545, Gerolamo Cardano published his ground-breaking book on algebra, *Ars Magna*, in which he investigated the solutions of some cubic equations. He introduced a new idea, that the square root of a negative number should be taken seriously and used in algebraic procedures rather than discarded as nonsense. In Rafael Bombelli's 1569 book *Algebra*, he developed the rules of complex algebra and pointed out that terms involving the square root of a negative number had to be segregated and treated separately from real numbers when performing addition or multiplication. Mathematicians subsequently realized the importance of $i$ as the key to a whole new branch of mathematics, the theory of functions of a complex variable, and this became an important part of the discipline of mathematics.

About 1750, Leonhard Euler used his skill with power-series expansions to introduce his famous formula: $\exp(i\theta) = \cos(\theta) + i\sin(\theta)$. The geometrical interpretation of this formula has found broad uses in physics and engineering. It makes the one-dimensional line of real numbers into a two-dimensional "complex plane" with the real numbers on the horizontal axis, the imaginary numbers on the vertical axis, with the variable $\theta$ representing the angle that a line from the origin to given point in the complex plane makes with the real axis. Entities like waves that have sinusoidal behavior can be very compactly described mathematically as rotations in the complex plane.

Electrical engineers often describe the time-dependent sinusoidal voltages on AC power lines as $V(t) = V_0 \sin(\omega t) = V_0 \Im\{\exp[i(\omega t)]\}$ where "$\Im\{\}$" means the imaginary part, i.e., using only the sin-function part of Euler's formula, $V_0$ is the



J.G. Cramer, *The Quantum Handshake*, DOI 10.1007/978-3-319-24642-0_4

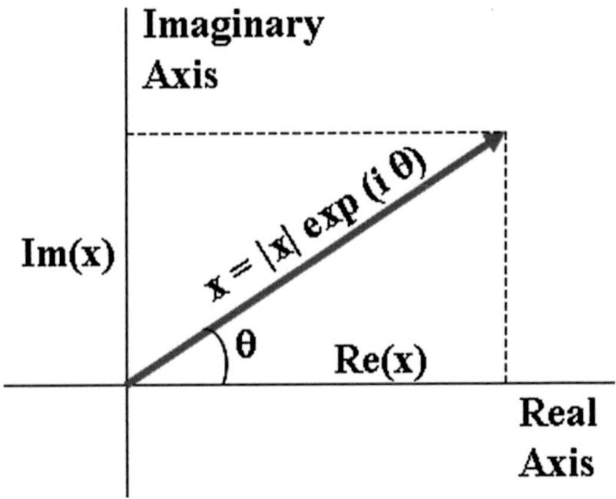


**Fig. 4.1** Representation of a complex variable $x = |x|\exp(i\theta)$ on the complex plane. The real part of $x$ projects on the horizontal axis, and the imaginary part of $x$ projects on the imaginary axis. The variable makes an angle $\theta$ with the real axis

peak voltage of the wave form, and $\omega$ is the angular frequency in radians per second of the voltage sine wave, i.e., the rate at which $\theta$ in Fig. 4.1 is changing with time. It is convenient, when engineers are dealing with combinations of such waves, to keep the whole exponential (not just the imaginary part) throughout a calculation, and to apply the extraction of the imaginary part only to the final result. We also note that electrical engineers typically use the symbol "$j$" rather than "$i$" to denote $\sqrt{-1}$ because "$i$" is already used by them as the symbol for electrical current.

In classical physics, a function describing a pressure wave or a displacement wave moving along the x-axis of a system, perhaps a sound wave or a wave on a string, is often written as $y(x,t) = y_0\cos(kx - \omega t) = y_0\,\Re\{\exp[i(kx - \omega t)]\}$. Here "$\Re\{\}$" means the real part of the function, $y_0$ is the amplitude of the displacement or pressure variation, and $k$ the wave number ($k = 2\pi/\lambda$), $\omega$ is the angular frequency ($\omega = 2\pi f$), $f$ is the frequency in Hz, and $\lambda$ is the wavelength of the wave.

In the wave functions of quantum mechanics, the extraction of the real part of the function is dropped and the whole complex function is used. Thus, the probability-amplitude wave function of a wave moving on the x-axis has the schematic form: $\psi(x,t) = \psi_0\exp(ikx - i\omega t)$. Quantum mechanics is unique among physics theories in "eating whole" the mathematical formalism of functions of a complex variable.

If we want a wave that moves in the negative space ($-x$) direction, we reverse the sign of the k-term to give $\psi_{x-rev}(x,t) = \psi_0\exp[-ikx - i\omega t]$. If we want a wave that moves in the negative time direction, we reverse the sign of the $\omega$-term to give $\psi_{t-rev}(x,t) = \psi_0\exp[ikx + i\omega t]$. The mathematical operation of complex conjugation, which replaces $i$ with $-i$ everywhere in a mathematical function, performs both of these reversals, so that $\psi^*(x,t) = \psi_0\exp[-ikx + i\omega t]$. In 1932, Eugene Wigner showed that in quantum mechanics, the operation of complex conjugation is the time-reversal operator, an operation that reverses the direction of time in the description of a quantum system.

Thus, the role of $i$ in quantum mechanics is a subtle one. Quantum wave functions are complex because their time structure is encoded in their complexity, and the operation of complex conjugation reverses that time structure, making time run backwards in the quantum mechanical description.

## 4.2 Dirac and Time Symmetry

Paul Dirac (Fig. 4.2) was among the originators of the standard formalism of quantum mechanics, developing a general formalism and showing that both Heisenberg's matrix mechanics and Schrödinger's wave mechanics were included in it [1]. In 1928, he introduced the Dirac equation, a wave equation compatible with relativity that is specific to "fermion" particles [2]. Fermions (examples: electrons, $\mu$-leptons, protons, quarks, ...) have an intrinsic angular momentum or "spin" that is a half-integer fraction of $\hbar$ (typically, $\hbar/2$) and obey the rules of Fermi-Dirac statistics, a generalization of the Pauli Exclusion Principle. In particular, fermions obey statistical rules that are qualitatively different from those observed by "boson" particles (examples: photons, $\alpha$-particles, $\pi$-mesons, gluons, ...), which have a spin of 0 or some integer

**Fig. 4.2** In 1933, Paul A.M. Dirac (1902–1984) received the Nobel Prize in Physics for his work on quantum mechanics

multiple of $\hbar$ and obey the statistical rules of Bose–Einstein statistics. It is interesting and puzzling to note that if bosons are rotated by 360° they return to the same state, while fermions must be rotated by 720° to return to the same state.

One of the unexpected predictions of the Dirac equation was the existence of anti-matter. Each fermion particle described by the Dirac equation should have a mirror-twin "antiparticle" that has the same mass and spin and an opposite electrical charge and parity. Dirac predicted that the negative-charge electron should have a positive-charge antimatter twin, the positron. In 1932, four years after Dirac's prediction, the positron was observed by Carl Anderson in cosmic-ray events in a cloud chamber.

But the Dirac equation raised other problems. In 1938, Dirac attempted to address one-such problem that had been troubling physics for many years, the "self-energy problem". Particles like electrons and positrons are described by the Dirac equation as having no structure, so in essence they are point-like spin-$^1/_2$ objects with an electric charge, a spin, and a dipole magnetic field.

However, electric and magnetic fields represent energy stored in space, with the energy in a given volume of space proportional to the square of the field strength in that region. If we imagine approaching an electron and measuring its field as we approach it, we would find that the field grows as the inverse square of the diminishing distance. If we reduce our distance from the electron by a factor of $^1/_2$, the electric field should grow by a factor of 4 and the energy per unit volume contained in the field should grow by a factor of 16. If we take the limit of this process as our distance from the electron goes to zero (because it is a point with no outer surface), the field and energy density should become infinite. Therefore, using Einstein's relation between mass and energy ($m = E/c^2$), each electron should, by this argument, have an infinite rest mass. As we know, electrons do not have infinite mass, so something must be wrong with this argument.

This is called the self-energy problem, and it represented a puzzle for both classical physics and quantum mechanics. It was more serious in quantum mechanics, however. In classical physics one could treat the electron as a sphere of charge with a radius and an outer surface at which the electric field stopped growing and adjust that radius to give the known mass of the electron. In quantum mechanics one is not allowed to give objects arbitrary unobserved properties, so electrons must be treated as point-like objects with no structure.

In his 1938 paper, Dirac suggested that perhaps the classical problem of self-energy was being handled incorrectly [3]. He pointed out that Maxwell's wave equation for electromagnetism had not one but two independent solutions (see Sect. 1.1). One of these solutions is the commonly-used "retarded" solution describing radiation that requires a time-delay to travel a spatial distance. The other solution is commonly ignored. It is the "advanced" solution describing radiation that requires a negative time-delay to travel a spatial distance. It can be thought of as an incoming rather than an outgoing wave, or as a wave that is traveling *backwards in time*. He proposed an alternate time-symmetric formalism for classical electrodynamics that used solutions of Maxwell's wave equation that are half of the retarded solution added to half of the advanced solution. In other words, instead of choosing the retarded solution by invoking causality, he assumed an even-handed treatment of time with no preferred time direction built into the formalism. He applied this to a description of how the electron would move in response to a force from a pulse of an external electric field. Dirac found that this permitted the elimination of the self-energy problem and led to a point-like electron with a finite mass.

Eventually, the solution to the self-energy problem was supposedly solved in quantum mechanics in another way that involved using vacuum polarization, a concept also originated by Dirac. In this approach, the bare charge of the electron is shielded at small distances by virtual particles of the opposite charge emerging from the vacuum, thereby reducing the mass-energy of the electric field to the observed value.[1] Nevertheless, Dirac's foray into time-symmetric electrodynamics was valid physics, it provided the basis for later work, and it should perhaps be reexamined in view of contemporary cosmological issues.

## 4.3 Wheeler–Feynman Absorber Theory

Before the beginning of World War II, Richard Feynman had been John Wheeler's graduate student at Princeton, and they had followed up Dirac's work with their own brand of time-symmetric electrodynamics. During WWII and the Manhattan Project, Feynman (Fig. 4.3) had led the calculation group at Los Alamos and Wheeler (Fig. 4.4) had been deeply involved in the production of plutonium at the first major nuclear reactor in Richland, Washington. At the end of WWII in 1945, they jointly published their work on time-symmetric electrodynamics, which became known as Wheeler–Feynman Absorber Theory [4, 5].

The classical electrodynamics described by Wheeler and Feynman (WF) was intended to deal with the problem of the self-energy of the electron in an innovative way. Assuming the time symmetric formalism of Dirac [3] combined with the *ad hoc* assumption that an electron does not interact with its own field, WF was able to formally eliminate the self-energy term from their electrodynamics. But along with self energy, these assumptions also removed the well observed energy loss and recoil processes (i.e., radiative damping) arising from the interaction of the radiating electron with its own radiation field.

However, WF accounted for these well-known damping effects by allowing the emitting electron to interact with the advanced waves sent by other electrons that would ultimately, at some time future, absorb the retarded radiation. Thus the energy

[1]We note that solving the self-energy problem by invoking vacuum polarization brings with it some unwanted baggage: the theory leads to an overestimate of the energy content of the vacuum, otherwise known as the cosmological constant, by a factor of $10^{120}$.

**Fig. 4.3** In 1965, Richard P. Feynman (1918–1988) received the Nobel Prize in Physics for his work on quantum electrodynamics, in which he introduced Feynman diagrams like the one shown here

loss and recoil of the emitter were accounted for without having it interact with its own field. Moreover, the calculation succeeded in describing electrodynamic interactions in a completely time-symmetric way. Effectively, the retarded and advanced waves together did a "handshake" that arranged for the transfer of energy and momentum (Fig. 4.5). To account for the observed asymmetric dominance of retarded radiation, WF invoked the action of external boundary conditions arising from thermodynamics. They thus avoided resort to the usual *ad hoc* "causality" condition usually needed to eliminate the advanced radiation solutions.

Regrettably, the work of Wheeler and Feynman, while mathematically correct, proved to be an invalid way of dealing with self-energy. As Feynman [6] later pointed out, the self-interaction they eliminated is a necessary part of electrodynamics, needed, for example, to account for the Lamb shift. And it is relevant that the WF *ad hoc* assumption of non-interaction is not needed in their recoil calculations because, as later authors have pointed out [7, 8], the electron cannot undergo energy loss or recoil, which are intrinsically asymmetric-in-time processes, as a result of interacting with its own (or any other) locally time-symmetric field.

When the offending *ad hoc* assumption of non-interaction is removed from the WF formalism, what remains is a classical self-consistent and time-symmetric electrodynamics that cannot be used to deal with the problem of self energy. Further, this WF formalism is not particularly useful as an alternative method of calculating the electrodynamics of radiative processes because the mathematical description of radiation explicitly involves the interaction of the emitter with the entire future universe.

**Fig. 4.4** John A. Wheeler (1911–2008) made significant theoretical contributions in nuclear physics and gravitational theory and originated the terms "black hole" and "wormhole"

Thus, a simple integration over local coordinates in the conventional formalism is replaced by an integral over all future space-time in the light cone of the emitter in the WF formalism.

However, this "difficulty" can be viewed an asset. The WF mathematics has been used to investigate the properties of cosmological models describing the future state of the universe by relating such models to radiative processes. In essence this approach provides a way of linking the cosmological arrow of time (the time direction in which the universe expands) to the electromagnetic arrow of time (the complete dominance of retarded over advanced radiation in all radiative processes). There is a considerable literature in this field, which the author has reviewed in a previous publication [9].

Although the original WF work dealt exclusively with classical electrodynamics, later authors [10–14] have developed equivalent time-symmetric quantum-electrodynamics (QED) versions of the same approach. The predictions of time-symmetric QED theories have been shown to be completely consistent with those predictions of conventional QED that can be compared with experimental observation. It has also been shown [14] that despite this similarity of prediction, the time-symmetric QED provides a qualitatively different description of electromagnetic processes.

**Fig. 4.5** The Wheeler–Feynman handshake, the basis of WF absorber theory and the Transactional Interpretation

It is essentially an action-at-a-distance theory with no extra degrees of freedom for the radiation fields and no second quantization. The field in effect becomes a mathematical convenience for describing action-at-a-distance processes.

There may also be another advantage to the WF approach to electrodynamics. Dirac's work [3] on time-symmetric electrodynamics, on which the WF theory is based, was introduced as a way of dealing with singularities in the radiation field in the conventional theory near a radiating electron. Emil Konopinski [15] in his Lorentz covariant treatment of the radiating electron has pointed out that this time-symmetric "Lorentz-Dirac" approach eliminates such singularities and therefore amounts to a self-renormalizing theory. This formulation may have applications in eliminating related singularities in QCD and in quantum field theory in curved space-time.

Richard Feynman's work on this problem led him to the formulation of his own brand of quantum mechanics, a framework using Lagrangians and action as starting points, instead of the more conventional approach using Hamiltonians and energy. This resulted in the application of Feynman path integrals to quantum mechanics and QED.

## References

1. P.A.M. Dirac, *The Principles of Quantum Mechanics* (Oxford University Press, London, 1930)
2. P.A.M. Dirac, The quantum theory of the electron. Proc. R. Soc. Lond. Series A **117**, 610–624 (1928)
3. P.A.M. Dirac, Proc. R. Soc. Lond. **A167**, 148 (1938)
4. J.A. Wheeler, R.P. Feynman, Rev. Mod. Phys. **17**, 157 (1945)
5. J.A. Wheeler, R.P. Feynman, Rev. Mod. Phys. **21**, 425 (1949)
6. R.P. Feynman, Phys. Rev. **76**, 769 (1949)
7. D.T. Pegg, Rep. Prog. Phys. **38**, 1339 (1975)
8. J.G. Cramer, Phys. Rev. D **22**, 362–376 (1980)
9. J.G. Cramer, Found. Phys. **13**, 887 (1983)
10. F. Hoyle, J.V. Narlikar, Ann. Phys. **54**, 207 (1969)
11. F. Hoyle, J.V. Narlikar, Ann. Phys. **62**, 44 (1971)
12. P.C.W. Davies, Proc. Camb. Philos. Soc. **68**, 751 (1970)
13. P.C.W. Davies, J. Phys. A **4**, 836 (1971)
14. P.C.W. Davies, J. Phys. A **5**, 1025 (1972)
15. E.J. Konopinski, *Electromagnetic Fields and Relativistic Particles* (Chapter XIV) (McGraw-Hill, New York, 1980)

# Chapter 5
# The Transactional Interpretation

## 5.1 Interpretations and Paradoxes

Let us start by considering again just what an "interpretation" of quantum mechanics is and why this very fundamental theory of physics needs an interpretation at all. As observed in Chap. 1, the theories of physics usually come with the interpretation as part of the package. Quantum mechanics is unique in having been formulated with a somewhat mysterious formalism that had no apparent interpretation. The Copenhagen Interpretation had to be rather painfully constructed by Bohr, Heisenberg, and Born in the late 1920s after the wave-mechanics and matrix-mechanics formalisms were already in place, in order to make calculations and predictions possible and to deflect some of the paradoxical messages (state vector collapse, simple nonlocality, wave-particle duality) that seemed to be implicit in the formalism.

For a simpler example of the relation of an interpretation to a theory, consider the formalism of Newton's Second Law: $F = ma$. If we had no knowledge of the meaning of the three symbols in the equation, the formalism would be a useless scribble. To use the equation, we must understand that $F$ symbolizes a force, and we must have some idea of what a force is, for example by thinking about the weight of an object in a gravitational field or the push we might give an object to move it across a table. We must understand that $m$ symbolizes mass and have some idea of mass related to the size and density of objects and the relative difficulty we might encounter in lifting or moving them.

And finally, we must understand that $a$ symbolizes acceleration. This is a more abstract concept than the other two, because it is the rate of change of the rate of change of the position of an object, an elusive concept that it took the insight of Newton to penetrate. We say, using the language of differential calculus, that acceleration is the second time-derivative of position. Knowing the meanings of the symbols then allows us to do calculations. We can predict the acceleration, given values for the mass and force. We can predict the force, given values for the mass



J.G. Cramer, *The Quantum Handshake*, DOI 10.1007/978-3-319-24642-0_5

and acceleration. Or if we are given values for the force and the acceleration we can calculate the object's mass.

But there are also some implicit paradoxes in the equation. Suppose the mass $m$ was zero. Then any applied force would produce an infinite acceleration. Or suppose the mass was negative. This would result in the paradoxical behavior that applying a force to the right would produce an acceleration to the left, the opposite of the direction that our experience leads us to expect. Or suppose the object resisted the action of a force more vigorously in some directions than in others, like a bead on a wire. Then the mass, instead of just having a value that was a simple number, would be a direction-dependent quantity, leading to many complications. And so, we must interpret Newton's second law as applying only to objects that have a positive non-zero mass, we must interpret mass as a direction-independent scalar, and we must assert that objects with negative or zero mass should not exist in our universe.[1]

These are the elements of the interpretation of Newton's Second Law. We do not usually separate this formalism from its interpretation and consider them separately, but it should be clear that they come together in the same package when we consider Newtonian mechanics.

The lesson of this example is that the interpretation of a formalism should serve two separate functions. It should interpret the symbols of the formalism so that we can perform calculations and make predictions, and it should provide intellectual tools that allow us to deal with seeming paradoxes implicit in the formalism. Further, if paradoxes are present, the interpretation needs to deal with *all* of them rather than focusing exclusively on some particular subset of what is perceived to be paradoxical implications of the formalism. We also note that the concept of a "paradox" is not completely well defined, and that one person's intellectual problem or challenge is another person's paradox.

As we have seen in Sect. 2.6, the Copenhagen Interpretation manages to fulfill the above criteria. It identifies those parts of the quantum mechanical formalism that are connected to experimental observation, and it provides detailed guidance on how to use the formalism to make calculations and predict the results of observations, at least in a probabilistic way. It deals with the internal mechanisms described by the formalism mainly with a positivist "don't-ask; don't-tell" approach, admonishing us to focus on measurement and observation and to shun inquiries about elements of the formalism that are not directly observable.

The Copenhagen Interpretation deals with paradoxes primarily with Heisenberg's knowledge interpretation, describing the quantum state vector as a mathematical representation of the state of knowledge of an observer of the system of interest, changing as measurements are made and new knowledge becomes available. There is no explanation or justification of how it could be that the solutions of simple second-order differential equations relating mass, momentum, and energy could conceivably be connected to observer knowledge. That's just the way it is. The resulting

[1] Particles like photons and gluons do have zero *rest* mass, but their dynamic mass has a non-zero value given by their net energy $E$ as $m = E/c^2$.

Copenhagen Interpretation is "air-tight", in that it deals with the procedures for the use of the formalism without providing any visualization tools for what might be going on behind the scenes.

A major part of the problem with understanding the role of quantum interpretations is that textbooks used to teach quantum mechanics at the graduate level often ignore the whole concept of an interpretation and by example encourage the graduate student to "shut up and calculate". I have surveyed twelve of the currently-used textbooks on quantum mechanics and have discovered the following statistics, based on examining the index pages of the books. While all 12 of these textbooks mention Heisenberg's uncertainty principle, only 5 out of the 12 of them actually mention the Copenhagen Interpretation [1–5]. Only 3 of 12 mention the Born rule [2–4], and only 2 of 12 mention Bohr's complementarity principle [3, 6]. None explicitly mention Heisenberg's knowledge interpretation. Only 2 of 12 mention nonlocality [2, 7], while 7 out of 12 mention entanglement [1–5, 7, 8]. Only 7 of 12 mention Bell's Theorem [1–5, 7, 8]. Only 6 of 12 mention the Einstein, Podolsky and Rosen work [1–5, 8]. And interestingly enough, 3 of the 12 mention the Many-Worlds interpretation [2–4] but none mention the Transactional Interpretation. Four textbooks, one of which shows a picture of Schrödinger's cat on the cover, do not mention any of these ideas except the uncertainty principle [9–12]. Thus, students learning quantum mechanics in graduate school are for the most part unaware that there are any interpretational problems, and they may or may not even be explicitly exposed to the ideas of the Copenhagen Interpretation.

## 5.2 The One-Dimensional Transaction Model

The starting point for the Transactional Interpretation of quantum mechanics is to view the "normal" wave functions $\psi$ appearing in the wave-mechanics formalism as Wheeler–Feynman retarded waves, to view the complex-conjugated wave functions $\psi^*$ as Wheeler–Feynman advanced waves, and to view the calculations in which they appear together in expectation-value integrals and matrix elements, as Wheeler–Feynman "handshakes" between emitter and absorber [13].

While there are notable similarities between the Wheeler–Feynman time-symmetric approach to electrodynamics and this approach to the quantum formalism, there are also important differences. In the classical electrodynamics of Wheeler–Feynman, it is the advanced-wave responses from *all* of the absorbers in the future universe, arriving together back at the emitter that cause it to radiate, lose energy, and recoil during emission. There are no photons and there is no quantization of energy, and so there is no single future absorber that receives all of the energy and momentum that the emitter has transmitted. Further, the emitter is responding to the full intensity of the superimposed advanced-wave fields from the future in a completely deterministic way, losing energy and gaining recoil momentum as a moving electric charge responding to external electric and magnetic fields.

In the domain of quantum mechanics these rules must be changed to reflect quantization and the probabilistic nature of quantum behavior. In the case of photon emission and absorption, an emitter emits a single photon, losing a quantum $-\hbar\omega$ of energy and experiencing momentum recoil $-\hbar k$. An absorber receives a single photon, gaining a quantum $\hbar\omega$ of energy and experiencing momentum recoil $\hbar k$. The rest of the future universe does not explicitly participate in the process. If a spherical wave function $\psi$ propagates for a significant distance before absorption, it becomes progressively weaker due to the inverse-square law, much too weak to be consistent with the behavior of an electric charge simply responding to external fields as in classical Wheeler–Feynman electrodynamics.

As an intermediate conceptual step, it is useful to think about the quantum situation in a single space dimension $x$ and in one time dimension $t$, so that the attenuation of the wave function with distance as it spreads out in three-dimensional space, which is a complication, can be put aside for the moment. This is then a wave-on-a-string situation in which the light cone becomes a diagonal Minkowski line connecting emitter to absorber, as shown in Fig. 5.1. In the spirit of even-handed time symmetry, the emitter must simultaneously send out retarded wave function $F_1(x,t) = \psi = A\exp[i(kx - \omega t)]$ and advanced wave function $G_1(x,t) = \psi^* = -A\exp[-i(kx - \omega t)]$ in the two time and space directions, i.e., in both directions from the emitter along the Minkowski line. The energy and momentum eigenvalues of $F_1$ are $\hbar\omega$ and $\hbar k$, while the eigenvalues of $G_1$ are $-\hbar\omega$ and $-\hbar k$. Therefore, the emission of the composite wave function $F_1 + G_1$ involves no change in energy or momentum, i.e., it has no energy or momentum cost. This is to be expected, since the emission process is time-symmetric, and time-symmetric fields should not produce any time-asymmetric loss of energy or momentum.

The absorber at some later time receives the retarded wave $F_1$ and terminates it by producing a canceling wave $F_2 = -A\exp[i(kx - \omega t)]$, as shown in Fig. 5.2.

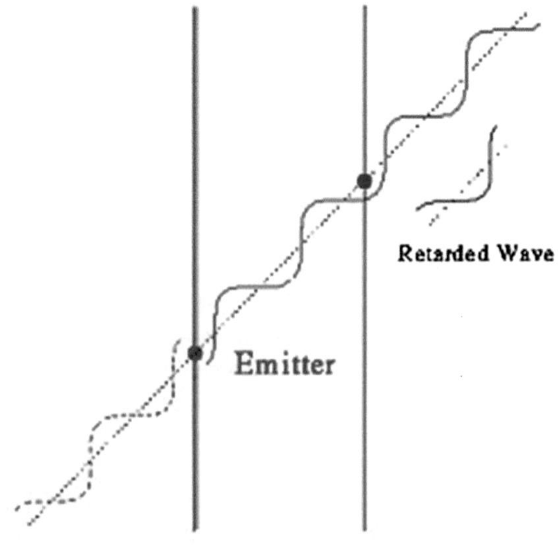


**Fig. 5.1** Schematic of the emission stage of a 1-dimensional transaction. An emitter (*red/left*) produces a retarded wave (*solid*) toward an absorber (*blue/right*) and an advanced wave (*dashed*) in the other time direction. Time is vertical and space horizontal

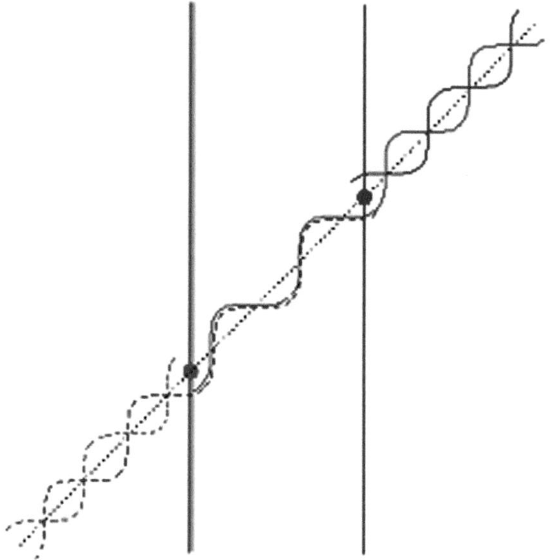

**Fig. 5.2** Schematic of the confirmation stage of a 1-dimensional transaction. An absorber (*blue/right*) responds with an advance wave (*dashed*) back to the emitter (*red/left*) and a retarded wave (*solid*) going forward in time beyond the absorber

Because the absorber must respond in a time-symmetric way, it must also produce advance wave $G_2 = A\exp[-i(kx - \omega t)]$, which travels back along the Minkowski line until it reaches the emitter. At the emitter it exactly cancels the advanced wave $G_1$ that the emitter had produced in the negative time direction.

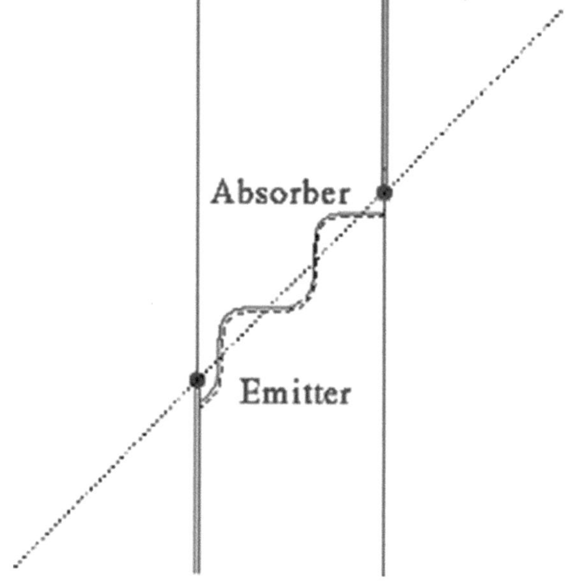


**Fig. 5.3** Schematic of the completed 1-dimensional transaction. Extra waves cancel, leaving an advanced-retarded "handshake" that transfers energy $\hbar\omega$ and momentum $\hbar k$ from emitter (*red/left*) to absorber (*blue/right*)

The net result is that a superposition of $F_1 + G_2$ connects emitter with absorber, the emitting charge interacts with $G_2$ by losing energy $\hbar\omega$ and recoiling with momentum $-\hbar k$, and the absorbing charge interacts with $F_1$ by gaining energy $\hbar\omega$ and recoiling with momentum $\hbar k$. Due to the cancellations beyond the interaction points $F_1 + F_2 = 0$ and $G_1 + G_2 = 0$, so there is no net wave function on the Minkowski diagonal before emission or after absorption. A Wheeler–Feynman handshake, shown in Fig. 5.3, has moved a quantum of energy $\hbar\omega$ and momentum $\hbar k$ from emitter to absorber. An observer, unaware of the time-symmetric processes involved, would say that a forward-going wave was emitted and subsequently absorbed.

## 5.3 The Three-Dimensional Transaction Model

Now let us consider the more realistic situation of three spatial dimensions and one time dimension. Assuming symmetric emission, the wave function $\psi$ spreads out in three dimensions, as shown in Fig. 5.4, like a bubble expanding from the central source location. The wave function, attenuated by distance, can reach many potential absorbers, each of which can respond by producing an advanced wave function $\psi^*$ that, also attenuated by distance, travels back to the emitter, as shown in Fig. 5.5. The emitter at the instant of emission can thus receive many advanced-wave "echoes". In this way, guided by the quantum formalism, attenuation and competition have been added to this picture.

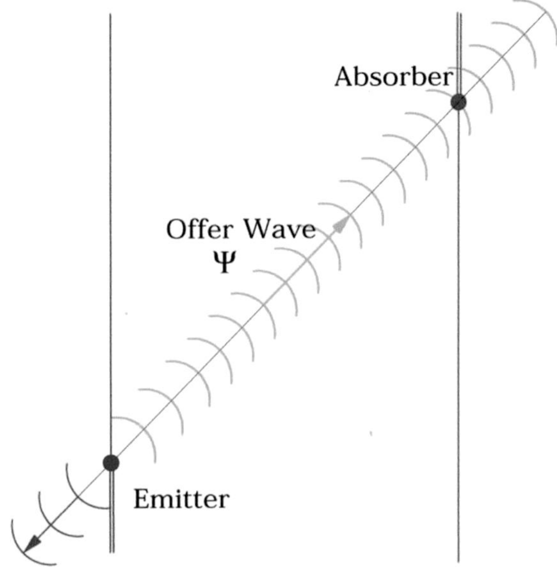


**Fig. 5.4** Schematic of offer wave $\psi$

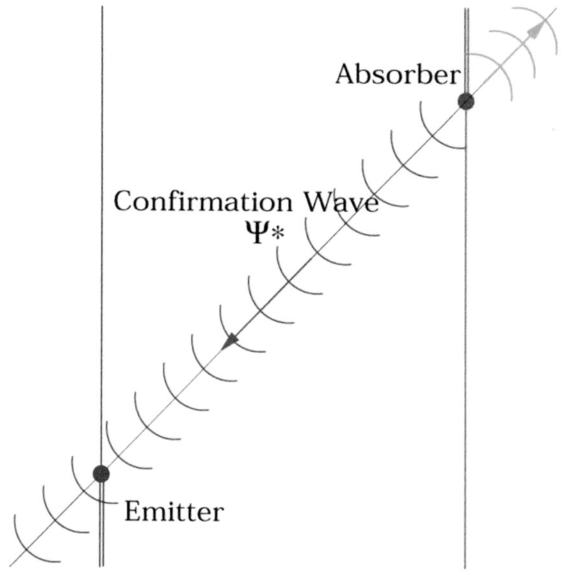


**Fig. 5.5** Schematic of confirmation wave $\psi^*$ response

We can think of the retarded waves from the emitter as offers to transmit energy or "offer waves", the first step in a handshake process that may ultimately produce the emission of a photon. Similarly, the advanced wave responses from potential absorbers can be thought of as "confirmation waves", the second step in the handshake process to transfer a photon. The advanced waves travel in the negative time direction and arrive back at the emission space-time location *at the instant of emission*, each with a strength $\psi$ that reflects the inverse-square-law attenuation of the offer wave in traveling forward from emitter to absorber and the reception strength of the absorber, multiplied by a strength $\psi^*$ that reflects the same attenuations of the confirmation wave in traveling back from absorber to emitter..

Therefore, the emitter receives an "echo" of magnitude $\psi_i \psi_i *$ from the $i$th potential future absorber. To proceed with the process, the emitter must "choose" one (or none) of these offer-confirmation echoes as the initial basis for a photon-emission handshake or "transaction", with the choice weighted in probability by the strength of each echo. If conserved quantities are involved in the transactions, the echos inconsistent with conservation are rejected. After the choice is made, there must be repeated emitter-absorber wave exchanges that grow exponentially until the strength of the space-time standing wave that thus develops is sufficient in strength to transfer a quantum of energy $\hbar\omega$ and momentum $\hbar k$ from the emitter to the absorber, completing the transaction, as shown in Fig. 5.6.

As a criticism of this transaction model, it might be argued that while the quantum wave function $\psi$ is a solution of the Schrödinger wave equation, its complex conjugate $\psi^*$ is not, and therefore the transaction model is inappropriately mixing solutions with non-solutions. However, we observe that the Schrödinger wave equation is inconsistent with Lorenz invariance and can be regarded as only the

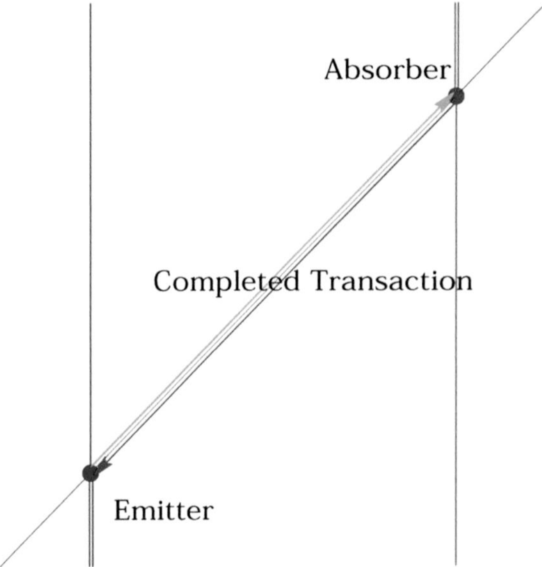


**Fig. 5.6** Schematic of the completed transaction as a standing wave in space-time

non-relativistic limit of the "true" relativistic wave equation, i.e., the Klein-Gordon equation for bosons or the Dirac wave equation for fermions, both consistent with relativity. Taking the non-relativistic limit of the Klein-Gordon or Dirac wave equation produces *two* wave equations, the Schrödinger wave equation and its complex conjugate. The wave function $\psi$ is a solution of the Schrödinger wave equation, while $\psi^*$ is a solution of the complex conjugate of the Schrödinger wave equation, and so both are equally valid solutions. The quantum version of the electromagnetic wave equation, which is relativistically invariant and is appropriate for describing the emission and absorption of photons, has both advanced and retarded solutions.

We note here that the sequence of stages in the emitter-absorber transaction presented here employs the use of "pseudo-time", describing a process between emitter and absorber as extending across lightlike or timelike intervals of space-time as a layered time sequence external to the process. This is motivated by the sequence of causal links in the process, but it is also a pedagogical device used for description. The process itself is atemporal, and the only observables come from the overall superposition of all of the steps that form the final transaction, which is essentially an advanced + retarded standing wave across space-time, connecting emitter and absorber. The pseudo-time and standing-wave accounts are complementary descriptions of the same process.

This is the transaction model by which the Transactional Interpretation describes the elements of the wave-mechanics formalism and accounts for quantum mechanical processes. The wave functions $\psi$ of the wave-mechanics formalism are the offer waves. In some sense they are real waves traveling through space, but in another sense they are not real because they represent only a mathematical encoding of the *possibility* of a quantum process. The transaction that forms after the emitter-absorber offer-confirmation exchange process goes to completion is the real object, what we would call the "particle" that has been transferred from emitter to absorber. In most cases, it has used or "projected out" only some specific component of the initial offer wave.

In that sense, the real objects in our universe are waves, while particles are an illusion created by the boundary conditions that must be observed at the vertices of the wave-exchange transactions. Can particles, little balls of mass-energy moving along on precise trajectories along trails blazed by the offer wave (a feature of the Bohm-de Broglie nonlocal hidden variable interpretation) actually exist in the quantum picture? Our consideration of the origins of the uncertainty principle in Sect. 2.4 tells us that the answer to this question is a resounding *no*. Entities that are completely described by wave functions cannot have simultaneous precise values of variables that are Fourier pairs, as needed for such trajectories. "Hidden variables" are so extremely well hidden that they cannot exist at all.

What happens to the offer and confirmation waves that do not result in the formation of a transaction? Since the formation of a transaction produces all of the observable effects, such waves are ephemeral, in that they produce no observable effects, and their presence or absence has no physical consequences. Moreover, the field cancellations of classical Wheeler–Feynman electrodynamics [13, 14] apply to the first two stages of transaction formation (emission and confirmation), which would remove all of the loose ends of "orphan" offer and confirmation waves. However, in explaining seemingly paradoxical quantum phenomena such as interaction-free measurements [15, 16] or quantum computing, such ephemeral waves can be viewed as "feeling out" components of the system and providing non-classical information (see Sect. 6.13) even when no transaction forms.

The transactional model not only provides a description of the process that underlies the calculation of a quantum mechanical matrix element, but it also explains and justifies Born's probability rule [17]. In particular, it explains why a quantum event described by a wave function $\psi$ has a probability of occurrence given by $\psi\psi^*$. In the transaction model, the quantities $\psi\psi^*$ are the strengths of the advanced-wave echoes arriving back at the site of emission at the instant of emission. The "lightning strike" of a transaction formation depends probabilistically on the strengths of these echoes.

The Born probability rule is an assumption of the Copenhagen Interpretation, asserted axiomatically without justification as one of the tenets of the interpretation. On the other hand, the Born probability rule follows naturally from the transactional account of the Transactional Interpretation and does not need to be added as a separate assumption. In that sense, the Transactional Interpretation is superior to the Copenhagen Interpretation because it is more philosophically "economical", requiring fewer independent assumptions.

## 5.4 The Mechanism of Transaction Formation

Some critics of the Transactional Interpretation have asked why it does not provide a detailed mathematical description of transaction formation. This question betrays a fundamental misunderstanding of what an interpretation of quantum mechanics actually is. In our view, the mathematics is (and should be) exclusively contained in the standard quantum formalism itself. The function of the interpretation is to *interpret that mathematics*, not to introduce any new additional mathematics.[2]

It is true that the standard formalism of quantum mechanics does not contain mathematics that explicitly describes wave function collapse (which the TI interprets as transaction formation). However, there has been an application of the standard QM formalism in the literature that provides a detailed mathematical description of the "quantum-jump" exponential build-up of a transaction involving the transfer of a photon from one atom to another. Carver Mead does this in Sect. 5.4 of his book *Collective Electrodynamics* [18]. Briefly, Mead considers an emitter atom in an excited state with excitation energy $E_1$ and a space-antisymmetric wave function of $\psi_E = A_E(r)\exp(-iE_1t/\hbar)$ and a structurally-identical absorber atom in its ground state with excitation energy $E_0$ and a space-symmetric wave function of $\psi_A = S_A(r)\exp(-iE_0t/\hbar)$, where $A$ is an antisymmetric function and $S$ is a symmetric function. Both of these are stable states with no initial dipole moments.

He assumes that the initial positive-energy offer wave from the excited emitter atom $E$ interacting with the absorber atom $A$ perturbs it into a mixed state that adds a very small component of excited-state wave function to its ground-state wave function. Similarly, the negative-energy confirmation wave echo from the absorber atom interacting with the emitter atom perturbs it into a mixed state that adds a very small component of ground-state wave function to its excited-state wave function, as shown schematically in Fig. 5.7. Because of these perturbations, both atoms develop small time-dependent dipole moments that, because of the mixed-energy states, oscillate with a beat frequency $\omega = (E_1 - E_0)/\hbar$ and act as coupled dipole resonators. The phasing of their resulting waves is such that energy is transferred from emitter to absorber at a rate that initially rises exponentially. To quote Mead:

> The energy transferred from one atom to another causes an increase in the minority state of the superposition, thus increasing the dipole moment of both states and increasing the coupling and, hence, the rate of energy transfer. This self-reinforcing behavior gives the transition its initial exponential character.

In other words, he shows mathematically that the perturbations induced by the initial offer/confirmation exchange triggers the formation of a full-blown transaction in which a photon-worth of energy $E_1 - E_0$ is transferred from emitter to absorber.

Thus, mutual offer/confirmation perturbations of emitter and absorber acting on each other create a frequency-matched pair of dipole resonators as mixed states, and

[2]This principle, however, is violated by the Bohm-de Broglie interpretation (see Sect. 5.3), the Ghirardi-Rimini-Weber interpretation, and many other so-called interpretations. In that sense, these are not interpretations of standard quantum mechanics at all, but are rather alternative theories.

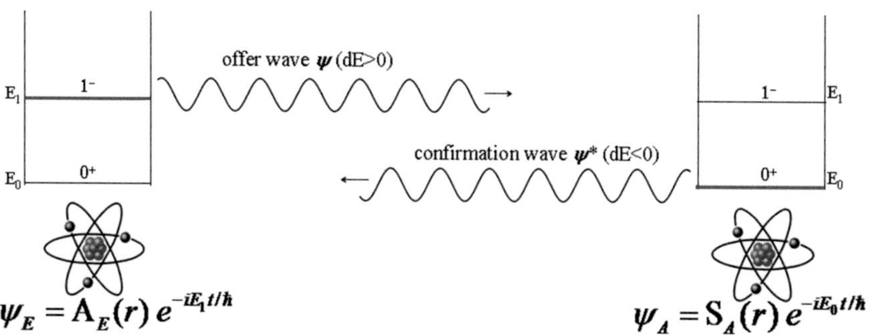


**Fig. 5.7** Mead model of transaction formation: Emitter in antisymmertric excited state of energy $E_1$ perturbs absorber in symmetric ground state of energy $E_0$ with offer wave, resulting in confirmation wave from absorber similarly perturbing emitter. Result is a pair of dipole resonators oscillating at beat frequency $\omega = (E_1 - E_0)/\hbar$, with exponentially rising coupling

this dynamically unstable system either exponentially avalanches to the formation of a complete transaction, or it disappears when a competing transactions forms. In a universe full of particles, this process does not occur in isolation, and both emitter and absorber are also randomly perturbed by waves from other systems that can randomly drive the exponential instability in either direction. This is the source of randomness in quantum processes. Ruth Kastner [19] likes to describe this intrinsic randomness as "spontaneous symmetry breaking", which perhaps clarifies the process by analogy with quantum field theory.

Because the waves carrying positive energy from emitter to absorber are retarded waves with positive transit time and the waves carrying negative energy from absorber to emitter are advanced waves with negative transit time, there is no net time delay, aside from time-of-flight propagation time of the transferred energy, in the quantum-jump process, and it is effectively instantaneous. Thus, the Transactional Interpretation explains Niels Bohr's "instantaneous" quantum jumps, a concept that Schrödinger found impossible to accept [20].

This is, of course, not a general proof that the offer/confirmation exchange always triggers the formation of a transaction, but it represents a demonstration of that behavior in a tractable case, and it represents a prototype of the general transaction behavior. It further demonstrates that the transaction model is implicit in and consistent with the standard quantum formalism, and it demonstrates how the transaction, as a space-time standing wave connecting emitter to absorber, can form.

## 5.5 Hierarchy and Transaction Selection

There is one more element of the transaction model, *hierarchy*, which needs to be added in order to avoid transactional inconsistencies pointed out by Maudlin [21] (see Sect. 6.14). All advanced-wave echoes are not equal. Those propagating back to the emitter from small space-time separation intervals rank higher in the selection hierarchy than those propagating back to the emitter from large space-time separation intervals. Selecting a transaction from among the advanced-retarded wave echos, the emitter makes an ordered separate decision to select *or not select* each particular echo for transaction formation. According to the mathematical description of transaction formation in Sect. 5.4 of reference [18], the echo perturbation triggers an exponential build-up to quantum transfer. If one echo does not trigger a transaction, the emitter then proceeds to make the same decision for the next closest echo, and so on. Thus, each echo is in competition with the decision to form no transaction and not normally in competition with the other echos. The emitter evaluates echos propagating back from small space-time intervals before proceeding to those echoes from larger intervals. This enforces the Born probability rule, but at the same time allows for the future to evolve, in part due to whether certain transactions were selected for completion or not.

The actual formation of a transaction, however, is likely to be more complicated than simple hierarchy would suggest. Lewis [22] has pointed out that a transaction involving two-path interference of massive particles, e.g. neutrons or electrons, would have echoes arriving with differing space-time intervals between emitter and absorber. Such transaction decisions would begin from the small-interval echo, but the completed transaction could not be competed until the responses on both paths were fully involved. In the case of destructive interference, the start of transaction formation might be terminated by canceling waves from the other path. In other words, as indicated by the Mead calculation, the formation of an transaction is a complicated process, and hierarchy is a useful simplification that does not tell the whole story.

Transaction formation hierarchy, simplification or not, has interesting implications for the structure of time itself in quantum processes. In some sense, the entire future of the universe may be reflected in the formation of each transaction, with the echoes from time-distant future events allowed the possibility of forming transactions only after the echoes from near-future absorbers have been weighed and rejected.

## 5.6 The Transactional Interpretation of Quantum Mechanics

The Transactional Interpretation of quantum mechanics [23–25], inspired by the structure of the quantum wave mechanics formalism itself, views each quantum event as a Wheeler–Feynman "handshake" or "transaction" process extending across space-time that involves the exchange of advanced and retarded quantum wave functions to enforce the conservation of certain quantities (energy, momentum, angular momentum, etc.). It asserts that each quantum transition forms in four stages: (1) emission, (2) response, (3) stochastic choice, and (4) repetition to completion.

The first stage of a quantum event is the emission of an "offer wave" by the "source", which is the object supplying the quantities transferred. The offer wave is the time-dependent retarded quantum wave function $\psi$, as used in standard quantum mechanics. It spreads through space-time until it encounters the "absorber", the object receiving the conserved quantities.

The second stage of a quantum event is the response to the offer wave by any potential absorber (there may be many in a given event). Such an absorber produces an advanced "confirmation wave" $\psi^*$, the complex conjugate of the quantum offer wave function $\psi$. The confirmation wave travels in the reverse time direction and arrives back to the source at precisely the instant of emission, delivering a retarded/advanced "echo" with an amplitude of $\psi\psi^*$. These echoes will not all be of the same strength, because they are attenuated by distance due to the inverse square law and may be of different magnitudes due to the characteristics of the absorber. For example, one absorber may have an electric quadrupole moment that produces the confirmation response while another may have a dipole moment.

The third stage of a quantum event is the stochastic choice exercised by the source in selecting one from among the possible transactions. It does this in a linear probabilistic way based on the strengths $\psi\psi^*$ of the advanced-wave "echoes" it receives from all of the potential absorbers. The selection is hierarchical (see Sect. 5.5), proceeding from echoes coming from the smallest space-time intervals of separation to echoes from larger separations. We note that one of the random choices available to the source is to form *no transaction at all*, an option that is particularly important in slow processes like $\alpha$-decay.

The final stage of a quantum event is the repetition to completion of this process by the source and the selected absorber, reinforcing the selected transaction repeatedly until the conserved quantities are transferred, the transaction is completed, and the potential quantum event becomes real.

Here we summarize the principal elements of the Transactional Interpretation, structured in order to contrast it with the Copenhagen Interpretation (see Sect. 2.6):

- The fundamental quantum mechanical interaction is taken to be the transaction. The state vector $\psi$ of the quantum mechanical formalism is a physical wave with spatial extent and is identical with the initial "offer wave" of the transaction. The complex conjugate of the state vector $\psi^*$ is also a physical wave and is identical with the subsequent "confirmation wave" of the transaction. The particle (photon, electron, etc.) and the collapsed state vector are identical with the completed transaction. The transaction may involve a single emitter and absorber and two vertices or multiple emitters and absorbers and many vertices, but is only complete when appropriate quantum boundary conditions are satisfied at all vertices, i.e., loci of emission and absorption. Particles transferred have no separate identity independent from the satisfaction of the boundary conditions at the vertices.
- The correspondence of "knowledge of the system" with the state vector $\psi$ is a fortuitous but deceptive consequence of transaction formation, in that such knowledge must necessarily follow and describe the transaction but does not cause it.

Thus, there is a confusion of cause with effect. Moreover, the knowledge interpretation predicts that the state vector, i.e., the offer wave, should change in mid-flight whenever new knowledge is obtained by measurement or by deduction and should disappear elsewhere when measurement is made at a particular location. This is not a part of the Transactional Interpretaion or of standard quantum mechanics. The user should avoid assuming that the state vector is a 1:1 map of knowledge.

- Heisenberg's uncertainty principle [26] is a consequence of the fact that a transaction in going to completion is able to project out and localize only one of a pair of conjugate variables (e.g., position or momentum) from the offer wave, and in the process it delocalizes the other member of the pair, as required by the mathematics of Fourier analysis. Thus, as in the Copenhagen Interpretation, the Uncertainty Principle is a consequence of the transactional model and is not a separate assumption.
- Born's probability rule [17] is a consequence of the fact that the magnitude of the "echo" received by the emitter, which initiates a transaction in a linear probabilistic way, has strength $P = \psi\psi^*$. Thus, Born's Probability Rule is a consequence of the transactional model and is not a separate assumption of the interpretation.
- All physical processes have equal status, with the observer, intelligent or otherwise, given no special status. Measurement and measuring apparatus have no special status, except that they happen to be processes that connect and provide information to observers.
- Bohr's "wholeness" of measurement and measured system exists, but is not related to any special character of measurements but rather to the connection between emitter and absorber through the transaction.
- Bohr's "complementarity" between conjugate variables exists, but like the uncertainty principle is just a manifestation of the requirement that a given transaction going to completion can project out only one of a pair of conjugate variables, as required by the mathematics of Fourier analysis.
- Resort to the positivism and "don't-ask-don't-tell" is unnecessary and undesirable. A distinction is made between observable and inferred quantities. The former are firm predictions of the overall theory and may be subjected to experimental verification. The latter, particularly those that are complex quantities, are not verifiable and are useful only for visualization, interpretational, and pedagogical purposes.

In summary, the Transactional Interpretation explains the origin of the major elements of the Copenhagen Interpretation while avoiding their paradoxical implications. It drops the positivism of the Copenhagen Interpretation as unnecessary, because the positivist curtain is no longer needed to hide the nonlocal backstage machinery.

It should also be pointed out that giving some level of objective reality to the state vector colors all of the other elements of the interpretation. Although in the Transactional Interpretation, the uncertainty principle and the statistical interpretation are formally the same as in the Copenhagen Interpretation, their philosophical implications, about which so much has been written from the Copenhagen viewpoint, may be quite different.

## 5.7 Do Wave Functions Exist in Real 3D Space or Only in Hilbert Space?

In classical wave mechanics, propagating waves, e.g. light or sound waves, are viewed as existing in and propagating through normal three-dimensional space. However, early in the development of the formalism of quantum mechanics it was realized that there was a problem with treating the quantum wave functions of multi-particle systems in the same way. Because of conservation laws and entanglement, the uncollapsed wave function of each particle in such a system was not only a function of its own space and momentum coordinates and other variables (e.g. spin and angular momentum), but might also be dependent on the equivalent coordinates of the other particles in the system. For example, the momentum magnitude and direction of each particle of a quantum system might be unspecified and allowed to take on any value over a wide range, but their momenta must be correlated so that the overall momentum of the system can have a well defined momentum value.

Therefore, it was concluded that Hilbert space provided a general way of describing quantum systems and that in multi-particle systems, a quantum mechanical wave function could not exist in simple three-dimensional space, but must instead reside in a higher Hilbert space of many more dimensions, with a dimension for each relevant variable, so that it was connected to all the variables needed for the description of its state. The wave function of a "free" independent non-entangled particle in such a space simply ignores the extra Hilbert space dimensions, allowing the extra variables to take on any value because there is no dependence on them. In such a Hilbert space the inter-dependences of multi-particle systems could be described, conservation laws could be defined as "allowed regions" that the wave functions could occupy, and powerful mathematical operations appropriate to higher dimensional spaces could be applied to the quantum formalism. The assertion that quantum wave functions cannot be considered to exist in normal space and must be viewed as existing only in an abstract higher-dimensional space, of course, creates a severe roadblock for any attempt to visualize quantum processes.[3]

The "standard" Transactional Interpretation described here, with its insights into the mechanism behind wave function collapse through transaction formation, provides a new view of the situation that make the retreat to Hilbert space unnecessary. The offer wave for each particle can be considered as the wave function of a free (i.e., uncorrelated) particle and can be viewed as existing in normal three dimensional space. The initial offer wave from a source is in an indefinite state (see the discussion of "realism" in Sect. 6.19).

It is conventional in the standard formalism to write for each particle only that projection of the solutions to the wave equation that will eventually fulfill the final conditions of entanglement, but this is just a convention that keeps the notation

[3] We note that Ruth Kastner's "Possibilist Transactional Interpretation" [19, 27] adopts just this point of view and treats quantum wave functions as being real objects only in an abstract multidimensional Hilbert space, from which transactions emerge in real space. The possibilist approach is perhaps not incorrect, but we consider it to be an unnecessarily abstract roadblock to visualization.

compact. A multitude of other solutions are there in principle in the initial wave function, but they are not explicitly denoted because they cannot form transactions due the conservation constraints, and so they never appear in a compact notation.

The application of conservation laws and the influence of the variables of the other particles of the system comes not in the offer wave stage of the process but in the formation of the transactions. The transactions "knit together" the various otherwise independent particle wave functions that span a wide range of possible parameter values into an interaction, and only those wave function sub-components that are correlated to satisfy the conservation law boundary conditions are permitted to participate in transaction formation.

Let's consider a concrete three-vertex three-particle example that shows how the transaction formation produces the correlations: a source of photons produces a polarizaton-entangled pair of photons that are subsequently absorbed by polarization-sensitive Detectors 1 and 2. In the standard formalism, we would write the source wave function, which is the initial offer wave, as:

$$| S\rangle = \alpha \mid H\rangle_1 \mid H\rangle_2 + \beta \mid V\rangle_1 \mid V\rangle_2 \tag{5.1}$$

where $H$ and $V$ represent horizontal and vertical polarization, respectively, $\alpha^2 + \beta^2 = 1$, and let us suppose for generality that $\alpha \neq \beta$. When Detector 1 receives the offer wave, it responds only to the $\mid H\rangle_1$ and $\mid V\rangle_1$ parts, which at this stage of the process are uncorrelated "free" wave functions, and returns confirmation wave echoes $|\alpha|^2\langle H \mid H\rangle_1$ and $|\beta|^2\langle V \mid V\rangle_1$ to the source. Similarly, Detector 2 responds only to the $\mid H\rangle_2$ and $\mid V\rangle_2$ parts and sends echoes $|\alpha|^2\langle H \mid H\rangle_2$ and $|\beta|^2\langle V \mid V\rangle_2$. The source, on receiving these echoes, has the choices of: (1) no response, (2) formation of a transaction with the $H$ polarizations, with relative probability $|\alpha|^2$, or (3) formation of a transaction with the $V$ polarizations, with relative probability $|\beta|^2$. It cannot form an $HV$ or $VH$ transaction because of the entanglement conditions. It is the source that chooses the form of the transaction, creates the correlations characteristic of the entangement, and makes sure that the conservation laws are observed. All of the waves involved are physically present, and in some sense "real", at least at the level of possibility, in normal three-dimensional space.

The "allowed zones" of Hilbert space arise from the action of transaction formation, not from constraints on the initial offer waves, i.e., particle wave functions. Hilbert space is the map, not the territory. Thus, the assertion that the quantum wave functions of individual particles in a multi-particle quantum system cannot exist in ordinary three-dimensional space is a misinterpretation of the role of Hilbert space, the application of conservation laws, and the origins of entanglement. Offer waves are somewhat ephemeral three-dimensional space objects, but only those components of the offer wave that satisfy conservation laws and entanglement criteria are permitted to be projected out in the final transaction, which, of course, also exists in three-dimensional space.

Another interesting question, relevant to the current need for a theory of quantum gravity, is whether the Transactional Interpretation would be consistent with the existence of a "universal" quantum wave function that described the state of the entire universe. The Copenhagen Interpretation, with its focus on observers and observer knowledge, has a severe problem with such a universal wave function. A universal wave function that would be interpreted as a description of observer knowledge would seem to require an observer *outside* the universe to have knowledge of it and to collapse its wave function. The Transactional Interpretation, which is independent of observers and observer knowledge, has no such problems. Further, it is relativistically invariant, and therefore should, in principle, be extensible to a theory of quantum gravity, should one emerge from the current effort.

## References


1. E. Merzbacher, *Quantum Mechanics*, 3rd edn. (Wiley, New York, 1977). ISBN: 978-0471887027
2. D.A.B. Miller, *Quantum Mechanics for Scientists and Engineers* (Cambridge University Press, New York, 2008). ISBN: 978-0521897839
3. G. Auletta, M. Fortunato, *Quantum Mechanics* (Cambridge University Press, New York, 2009). ISBN: 978-0521869638
4. S. Weinberg, *Lectures on Quantum Mechanics* (Cambridge University Press, New York, 2012). ISBN: 978-1107028722
5. J. Binney, D. Skinner, *The Physics of Quantum Mechanics* (Oxford University Press, Oxford, 2013). ISBN: 978-0199688579
6. N. Zettili, *Quantum Mechanics: Concepts and Applications* (Wiley, New York, 2009) ISBN: 978-0470026793
7. L. Susskind, A. Friedman, *Quantum Mechanics: The Theoretical Minimum* (Basic Books, New York, 2014) ISBN: 978-0465036677
8. E.D. Commins, *Quantum Mechanics: An Experimentalist's Approach* (Cambridge University Press, New York, 2014) ISBN: 978-1107063990
9. L.D. Landau, L.M. Lifshitz, *Quantum Mechanics, 3rd Ed.: Non-Relativistic Theory*, (Butterworth-Heinemann, 1981). ISBN: 978-0750635394
10. D.J. Griffiths, *Introduction to Quantum Mechanics* (Pearson, Upper Saddle River, 2004) ISBN: 978-0131118928
11. R. Shankar, *Principles of Quantum Mechanics*, 2nd edn. (Springer, New York, 2011) ISBN: 978-0306447907
12. D.G. Swanson, *Quantum Mechanics: Foundations and Applications* (Taylor & Francis, Abigton, UK, 2007). ISBN: 978-1584887522
13. J.A. Wheeler, R.P. Feynman, Rev. Mod. Phys. **17**, 219 (1945)
14. J.A. Wheeler, R.P. Feynman, Rev. Mod. Phys. **21**, 425 (1949)
15. A.C. Elitzur, L. Vaidman, Found. Phys. **23**, 987–997 (1993)
16. J.G. Cramer, Found. Phys. Lett. **19**, 63–73 (2006)
17. M. Born, W. Heisenberg, P. Jordan, Zur Quantenmechanik II. Z. für Phys. **35**(8-9), 557–615 (1926). (Third paper on the matrix mechanics formulation of quantum mechanics)
18. C. Mead, *Collective Electrodynamics* (The MIT Press, Cambridge, 2000) ISBN: 0-262-13378-4
19. R.E. Kastner, *The Transactional Interpretation of Quantum Mechanics: The Reality of Possibility* (Cambridge University Press, Cambridge, 2012)
20. W. Heisenberg, Reminiscences from 1926 and 1927. In Ref. [27]
21. T. Maudlin, *Quantum Nonlocality and Relativity* (Blackwell, Oxford) (1996, 1st ed.; 2002, 2nd. ed.)
22. P.J. Lewis, Retrocausal quantum mechanics: Maudlin's challenge revisited. Stud. Hist. Philos. Mod. Phys. **44**, 442–449 (2013)

23. J.G. Cramer, The transactional interpretation of quantum mechanics. Rev. Mod. Phys. **58**, 647–687 (1986)
24. J.G. Cramer, An overview of the transactional interpretation of quantum mechanics. Int. J. Theor. Phys. **27**, 227–236 (1988)
25. J.G. Cramer, The plane of the present and the new transactional paradigm of time, in (Chapter 9) *Time and the Instant*, ed. by R. Drurie (Clinamen Press, Manchester, 2001) arXiv:quant-ph/0507089
26. W. Heisenberg, Z. für Phys. **43**, 172 (1927), pp. 62–84 (translated in [28])
27. R.E. Kastner, On delayed choice and contingent absorber experiments. ISRN Math. Phys. **2012**, 617291 (2012)
28. A.P. French, P.J. Kennedy (eds.), *Niels Bohr, A Centernary Volume* (Harvard University Press, Cambridge, 1985)
29. J.A. Wheeler, W.H. Zurek (eds.), *Quantum Theory and Measurement* (Princeton University-Press, Princeton, 1983)

# Chapter 6
# Quantum Paradoxes and Applications of the TI

The Copenhagen Interpretation brings with it a certain baggage. In particular, Heisenberg's knowledge interpretation and positivism have led Einstein, Schrödinger, Wigner, Wheeler, and many others to focus on situations in which the conventional Copenhagen interpretational tools seem to fail, to lead to counter-intuitive conclusions, or to paradoxes. In this Chapter we will consider some of these, and we will also show examples applying the Transactional Interpretation to clarify quantum interpretational problems.

In the sections that follow, we will distinguish between *gedankenexperiments* that, for one reason or another have not been performed or cannot be performed, and actual experiments that have been performed and analyzed in the quantum optics laboratories by placing an asterisk (*) at the end of the section headings of the latter. The starred experiments have actually been performed.

## 6.1 Thomas Young's Two-Slit Experiment (1803)*

Thomas Young (1773–1829) presented the results of his two-slit experiment to the Royal Society of London on November 24, 1803. A century and a half later, Richard Feynman [1] described Young's experiment as "a phenomenon that is impossible ...to explain in any classical way, and that has in it the heart of quantum mechanics. In reality, it contains the only (quantum) mystery."

The experimental arrangement of Young's two-slit experiment is shown in Fig. 6.1. Plane waves of light diffract from a small aperture in screen $A$, pass through two slits in screen $B$, and produce an interference pattern in their overlap region on screen $C$. The interference pattern is caused by the arrival of light waves at screen $C$ from the two slits, with a variable relative phase because the relative path lengths of the two waves depends on the location on screen $C$. When the path lengths are equal



J.G. Cramer, *The Quantum Handshake*, DOI 10.1007/978-3-319-24642-0_6

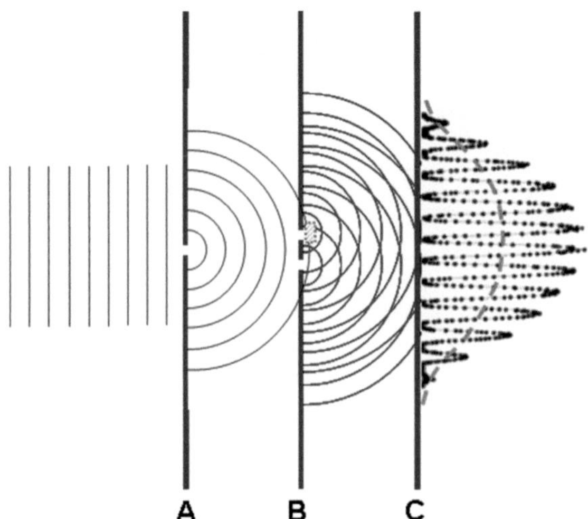


**Fig. 6.1** Young's two-slit experiment. Light waves diffract from the aperture in screen *A*, pass through two slits in screen *B*, and produce a "comb" interference pattern in their overlap region on screen *C*. The *green/dashed line* at *C* shows the diffraction pattern that would be observed if the two paths through the slits were made distinguishable, e.g., put in different states of polarization by a half-wave plate, shown behind the upper slit at *B*

or differ by an integer number of light wavelengths $\lambda$, the waves add coherently (constructive interference) to produce an intensity maximum. When the path lengths differ by an odd number of half-wavelengths $\lambda/2$, the waves subtract coherently to zero (destructive interference) and produce an intensity minimum.

One can "turn off" this interference pattern by making the two paths through slits *distinguishable*. In this case, the "comb" interference pattern is replaced by a broad diffraction "bump" distribution, as shown by the green/dashed line at *C* in Fig. 6.1. This might be accomplished by arranging for the waves on the two paths to be in different polarization states, thereby "labeling" the wave paths with polarization. For example, one could use a light source that produces vertically polarized light, and one could place behind one slit a small optical half-wave plate, shown in Fig. 6.1 behind the upper slit at *B*, set to rotate vertical to horizontal polarization. This would eliminate the previously observed two-slit interference pattern, because the light waves arriving at screen *C* from the two slits are now in distinguishable polarization states, with the waves from the lower slit vertically polarized and waves from the upper slit horizontally polarized. The *intensities* of the waves will now add instead of their amplitudes, and there can be no destructive cancellation. This interference suppression occurs even if no polarization is actually measured at *C*.

In the 19th century Young's experiment was taken as conclusive proof that light was a wave and that Newton's earlier depiction of light as a particle was incorrect. Einstein's 1905 explanation of the photoelectric effect as caused by the emission of photon particles of light cast doubt on this view. In 1909, a low-intensity double-slit experiment performed by Sir Geoffrey Taylor [2] demonstrated that the same interference pattern is obtained, even when the light intensity is so low that the interference pattern must be accumulated one photon at a time. The emergence of the interference pattern from individual photon events is illustrated in Fig. 6.2, in which we see the build-up of the two-slit interference pattern as single photon events (green points) are accumulated, one at a time. Based on Taylor's experimental results,

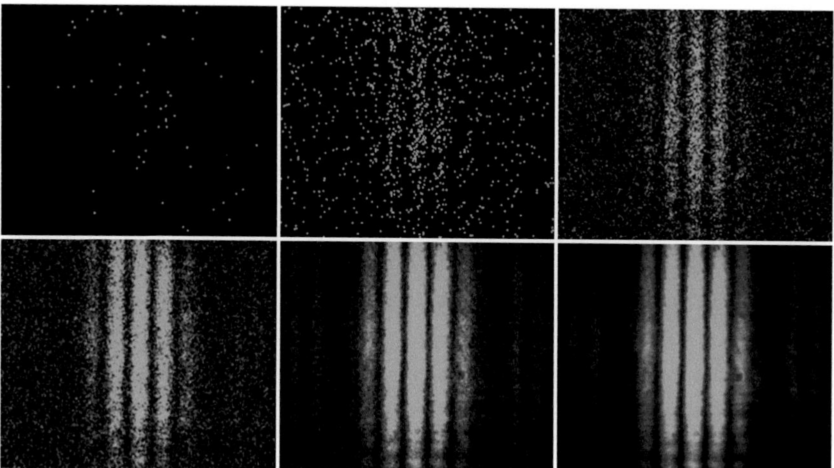

**Fig. 6.2** Build-up of a two-slit interference pattern in a Young's two-slit experiment at low illumination intensity as more and more single-photon events (*green points*) are accumulated [4]

in 1926 G.N. Lewis [3] reasoned, in a remarkable precursor to the Transactional Interpretation, that "an atom never emits light except to another atom …I propose to eliminate the idea of mere emission of light and substitute the idea of transmission, or a process of exchange of energy between two definite atoms or molecules."

The emergence of the interference pattern from individual photon events is the "quantum mystery" to which Richard Feynman referred: How is it possible that an ensemble of single photons, arriving at the screen one at a time, can produce such a wave-like interference pattern? It would appear that each individual photon particle must pass through *both* slits and must interfere with itself at the screen.

The Transactional Interpretation explains the puzzling build-up of a wave interference pattern from photon events as follows: in Fig. 6.1 the source emits plane offer waves moving to the right that are diffracted at screen *A*, pass through both slits at screen *B*, and arrive at any point on screen *C* from two directions. At locations along screen *C* where the two components of the offer wave interfere constructively there is a high probability of transaction, and at locations where the two components of the offer wave interfere destructively and cancel there is zero probability of a transaction. Confirmation waves propagate to the left, moving back through the slits at *B* and the aperture at *A* to the light source. There the source, which is seeking to emit one photon, selects among the confirmation offers, and a transaction delivers a photon to screen *C*. The position at which the photon arrives is likely to be where the offer waves were constructive and unlikely to be where the waves were destructive. Therefore, the interference pattern made of many single photon transactions builds up on screen *C* as shown in Fig. 6.2.

The interference suppression from labeling can also be explained by the TI. Screen *C* receives offer waves that have passed through both slits and returns corresponding confirmation waves to the source. However, the vertically polarized offer wave will cause the return of a vertically polarized confirmation, and likewise for the horizontally polarized offer wave. The confirmation wave echo arriving at the source will only match the vertical polarization of the source if it returned through the same slit that the corresponding offer had passed through, so the transaction that forms will pass through only one of the two slits. Therefore, there will be no two-slit interference pattern for this case.

## 6.2 Einstein's Bubble *Gedankenexperiment* (1927)

Quantum nonlocality is one of the principal counterintuitive aspects of quantum mechanics. Einstein's "spooky action-at-a-distance" is a real feature of quantum mechanics, but the quantum formalism and the orthodox Copenhagen Interpretation provide little assistance in understanding nonlocality or in visualizing what is going on in a nonlocal process. The Transactional Interpretation provides the tools for doing this. Perhaps the first example of a nonlocality paradox is the Einstein's bubble paradox, previously mentioned in Sect. 1.1. It was proposed by Albert Einstein at the 5th Solvay Conference in 1927 [5, 6].

A source emits a single photon isotropically, so that there is no preferred emission direction. According to the Copenhagen view of the quantum formalism, this should produce a spherical wave function $\psi$ that expands like an inflating bubble centered on the source. At some later time, the photon is detected, and, since the photon does not propagate further, its wave function bubble should "pop", disappearing instantaneously from all locations except the position of the detector. Einstein asked how the parts of the wave function away from the detector could "know" that they should disappear, and how it could be arranged that only a single photon was always detected when only one was emitted?

At the 5th Solvay Conference, Werner Heisenberg [6] dismissed Einstein's bubble paradox by asserting that the wave function cannot be depicted as a real object moving through space, as Einstein had implicitly assumed, but instead is a mathematical representation of the knowledge of some observer who is watching the process. Until detection, the observer knows nothing about the location of the emitted photon, so the wave function must be spherical, distributed over the $4\pi$ solid angle to represent his ignorance. However, after detection the location of the photon is known to the observer, so the wave function "collapses" and is localized at the detector. One photon is detected because only one photon was emitted.

The Transactional Interpretation provides an alternative explanation, one that permits the wave function to be, in some sense, a real object moving through space. This is illustrated in Fig. 6.3. The offer wave $\psi$ from the source indeed spreads out as a spherical wave front and eventually encounters the detector on the right. The detector responds by returning to the source a confirmation wave $\psi^*$. Other detectors (i.e., potential absorbers) also return confirmation waves, but the source randomly, weighted by the $\psi\psi^*$ echoes from the potential absorbers, selects the detector on the right to form a transaction. The transaction forms between source and detector, and

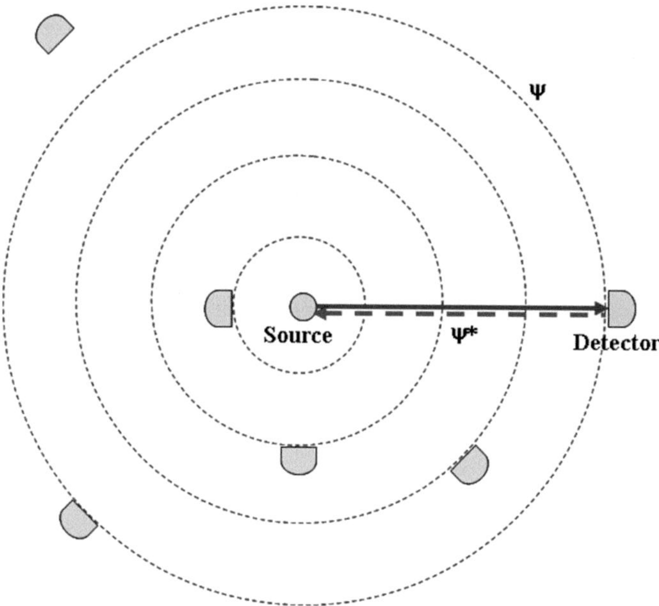


**Fig. 6.3** Schematic of the transaction involved in the Einstein's bubble paradox. The offer wave $\psi$ (*blue/solid*) forms a spherical wave front, reaching the detector on the right and causing it to return a confirmation wave $\psi^*$ (*red/dashed*), so that a transaction forms and one photon's worth of energy $\hbar\omega$ is transferred. Other detectors also return confirmation waves, but the source has randomly selected the detector on the right for the transaction

one $\hbar\omega$ photon's worth of energy is transferred from the source to the detector. The formation of this particular transaction, satisfying the source boundary condition that only one photon is emitted, prevents the formation of any other transaction to another possible photon absorber, so only one photon is detected. This is an illustration of a simple two-vertex transaction in which the transfer of a single photon is implemented nonlocally. It avoids Heisenberg's assertion that the mathematical solution to a simple second-order differential equation involving momentum, energy, time, and space has somehow become a map of the mind, deductions, and knowledge of a hypothetical observer.

In this context, we note that there is a significant (but untestable) difference between Heisenberg's knowledge interpretation and the Transactional Interpretation as to whether the outgoing state vector or offer wave changes, collapses, or disappears at the instant when knowledge from a measurement is obtained. The knowledge interpretation would lead us to expect, without any observational evidence and with some conflict with special relativity, that Einstein's bubble "pops" when the detector registers the arrival of a photon and that other parts of the outgoing wave disappear at that instant. The bubble needs to pop in the knowledge interpretation because the state of knowledge changes, and also because this prevent multiple photon detections from a single photon emission.

In the analogous description by the Transactional Interpretation, the parts of the offer wave away from the detection site, because they represent only the *possibility* of a quantum event, do not disappear, but instead continue to propagate to more distant potential detection sites. These sites return confirmation echoes that compete with the echo from the detector of interest for transaction formation. The consequence of this difference is that the TI does not have to explain how wave functions can change in mid-flight, how the *absence* of a detection can change a propagating wave function, or what "instantaneous" means in the context of special relativity. See the discussions of Renninger's and Maudlin's *gedankenexperiments* in Sects. 6.6 and 6.16 for further examples of this important interpretational difference.

## 6.3 Schrödinger's Cat (1935)

In 1935, Erwin Schrödinger presented his Cat Paradox, a problem that focused on the situation that occurs when the strange procedures of quantum mechanics in acting on microscopic systems are projected into the macroscopic world [7]. It is illustrated in Fig. 6.4. Suppose, he said, that we have a box that is completely and perfectly

(William R. Warren, Jr., © 1985, reproduced with permission.)

**Fig. 6.4** The Schrödinger's Cat *gedankenexperiment*: A sealed and insulated box (*A*) contains a radioactive source (*B*) that has a 50 % chance during the course of the"experiment" of triggering Geiger counter (*C*) that activates a mechanism (*D*) causing a hammer to smash a flask of hydrocyanic acid (*E*) killing the cat (*F*). In the Copenhagen view, the observer (*G*) must open the box in order to collapse the wave function of the system into one of the two possible states: $|\,\text{alive}\rangle$ or $|\,\text{dead}\rangle$, and before that, the cat's wave function was $(1/\sqrt{2})\,|\,\text{alive}\rangle + (1/\sqrt{2})\,|\,\text{dead}\rangle$

isolated from the outside world and has its own air supply. In the box is a radioactive source, a Geiger counter, and a mechanism that will smash a flask of hydrocyanic acid (a lethal poison) if the Geiger counter should detect a single radiation event. The radioactive source is very weak, with a strength adjusted so that the probability of the Geiger counter detecting a single radiation event in one hour is just 50 %.

Now we place a cat in the box, seal the lid, and wait for an hour. The question is, what is the quantum mechanical wave function describing the state of the cat at the end of an hour? There is a probability of $1/2$ that the cat will be alive at the end of an hour and a probability of $1/2$ that it will be dead. According to the procedures and formalism of standard quantum mechanics, the wave function of the cat is therefore $\Psi_{cat} = (1/\sqrt{2}) \mid \text{alive}\rangle + (1/\sqrt{2}) \mid \text{dead}\rangle$. In other words, the cat is predicted to be in a state that is half alive and half dead, two inconsistent states, as Schrödinger put it, that are mixed or smeared out in equal parts. He expressed an unwillingness to accept as valid such a "blurred model" for representing reality.

I have not been able to determine whether Heisenberg ever addressed the Schrödinger's Cat paradox directly, but his response is fairly predictable. The wave function of the cat is a mathematical representation of the knowledge of an observer. Since the observer does not know the state of the cat after an hour, of course a wave function representing his state of knowledge would have to include both dead and alive possibilities, and would be a mixture of the two until the box was opened.

The central focus of the problem posed by Schrödinger's Cat is the question of *when* the wave function actually collapses. The Transactional Interpretation avoids this implicit dilemma because in the TI the wave function collapse, i.e., the formation of the transaction, is two-way in time and atemporal. During the entire one hour period that the box is closed the radioactive source $B$ of Schrödinger's apparatus sends out a very weak offer wave $\psi$. This offer wave and its confirmation wave may or may not, with equal 50 % probabilities, be selected to produce a detection by the Geiger counter $C$, so that a completed transaction is formed. If a transaction is formed, then the count is recorded, the flask shattered, and the cat killed. If such a transaction is not formed then the cat remains alive. The initial wave function (or offer wave) does indeed have implicit in it both live cat and dead cat possibilities, but the completed transaction (or lack thereof) allows only one of these possibilities to become real. Because the collapse does not have to await the arrival of the observer, there is never a time when "the cat is 50 % alive and 50 % dead". And the need for consciousness, permanent records, thermodynamics, or alternate universes never arises. If the "buck stops" anywhere, it stops at the radioactive source at the start of the process, which receives advanced wave echoes from potential radiation absorbers and must select from among them transactions that can lead to only one of the two possible outcomes to be projected into reality, a live cat or a dead cat.

To put this another way, Schrödinger's question is: When can a quantum event be considered finished? Is it when the gamma ray leaves the radioactive nucleus? Is it when it interacts with the Geiger counter? When the flask is smashed? When the cat dies? When the observer looks in the box? When he tells a colleague what he observed? When he publishes his observations in the *Physical Review*? When …? A billiard shot is over when the billiard balls stop colliding and come to rest. But the atomic "billiard balls" of a quantum billiard game continue to collide forever, never coming to rest so that the shot can be considered finished.

The source of confusion here is that the wrong question is being asked. The Copenhagen view has led us to ask *when* the wave function collapses instead of *how* it collapses. But there is not a "when", not a point in time at which the quantum event is finished. The event is finished when the transaction forms, which happens along a set of world lines which include all of the event listed above, treating none of them as the special conclusion of the event. If there is one particular link in this event chain which is special, it is not the one which ends the chain. It is the link at the beginning of the chain when the emitter, having received various confirmation waves from its original offer wave, reinforces one (or more) of them in such a way that it brings that particular confirmation into reality as a completed transaction. The atemporal transaction does not have a "when" at the end.

## 6.4 Wigner's Friend (1962)

In 1962 Eugene Wigner elaborated on the knowledge issue with his Wigner's Friend paradox, an expansion of the Schrödinger's Cat problem [11]. Wigner replaced the cat with a "friend", i.e., an intelligent observer and at the same time replaced the hydrocyanic acid mechanism with a less lethal piece of apparatus, e.g., a light bulb that is switched on when a count is recorded. The experimenter then performs the experiment, which can be considered as two experiments: (a) treating friend+box as a system, the experimenter makes an observation, and (b) treating the counter mechanism as a system, the friend makes an observation that is subsequently reported to the experimenter (Fig. 6.5).

We will not reproduce Wigner's detailed analysis of this *gedankenexperiment* here, but will state his conclusion: consciousness must have a special role in the collapse of the wave function, for otherwise one must deal (at least on the philosophical level) with un-collapsed wave functions containing conscious observers in a multiplicity of alternative states.

The discussion in Sect. 6.3 also applies to the Wigner's Friend paradox. From the viewpoint of the Transactional Interpretation, there is nothing special about one observer observing another one. Transactions involving observers, like other transactions, form atemporally, and asking *when* the transaction forms is asking the wrong question.

**Fig. 6.5** The Wigner's Friend *gedankenexperiment*: In the Schrödinger's Cat setup, a second observer (Wigner's Friend *H*) may be needed, according to the Copenhagen view, to collapse the wave function of the larger system containing the first observer (*G*) and the apparatus (*A–F*). And another observer may be required to collapse *his* wave function, and so on …

## 6.5 Renninger's Negative-Result *Gedankenexperiment* (1953)

This is a *gedankenexperiment* focusing on the collapse of the wave function produced by the *absence* of an interaction of the system measured (an alpha-particle) with the measurement apparatus. It was suggested by Renninger [8] and was featured by de Broglie [9] in his book on the interpretation of quantum mechanics. The experimental arrangement is shown in Fig. 6.6.

Source $S$ is located at the center of a spherical shell $E_2$ of radius $R_2$. The interior of $E_2$ is lined with a scintillating material that will produce a detectable flash of light that will be seen by the observer if $E_2$ is struck by a charged particle, e.g., an alpha particle. Inside $E_2$ is a partial concentric sphere $E_1$ of radius $R_1$, also lined with scintillator viewed by the observer. Partial sphere $E_1$ subtends solid angle $\Omega_1$ as viewed from the position of source $S$. The portion of $E_2$ that is not shadowed by $E_1$ therefore subtends a solid angle $\Omega_2 = 4\pi - \Omega_1$. The source $S$ is arranged so that in the time interval of the experiment it will emit exactly one alpha particle with velocity $v$, which has an angular dependence that is completely isotropic.

A reminder about notation: in the discussions that follow we will explicitly indicate offer waves $\psi$ using the Dirac bra/ket state vector notation; a *ket* is a bar and angle bracket that enclose some symbol that distinguishes one retarded offer wave function

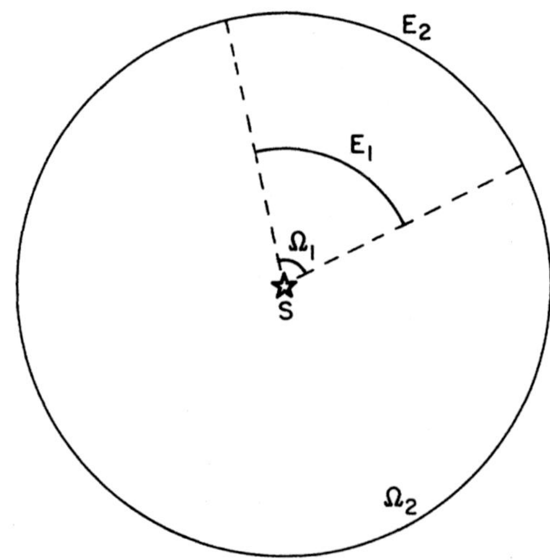


**Fig. 6.6** Schematic diagram showing Renninger's negative-result *Gedankenexperiment*. Source $S$ isotropically emits an alpha particle that is detected by the scintillator on spherical surfaces $E_1$ and $E_2$, depending on respective solid angles $\Omega_1$ and $\Omega_2$

from another. For example, a wave that is terminated at partial sphere $E_1$ can be represented by $\psi_{E_1} = \mid E_1\rangle$. The corresponding confirmation waves $\psi^*_{E_1}$ will similarly be indicated by a Dirac *bra* state vector $\psi^*_{E_1} = \langle E_1 \mid$, which reverses the bar and angle bracket to indicate an advanced confirmation wave.

Now let us consider the state vector $\mid S(t)\rangle$ as a function of time $t$, where $t$ is the time that has elapsed since the source $S$ has been commanded to emit an alpha particle. At time $t$ before the alpha particle has traversed the distance $R_1$, i.e., for $0 < t < (R_1/v)$, the probability that the particle will produce a scintillation at $E_1$ is $P_1 = \Omega_1/4\pi$, and the probability that it will produce a scintillation at $E_2$ is $P_2 = \Omega_2/4\pi$. Thus the state vector might be written as:

$$\mid S(t)\rangle = p_1 \mid E_1\rangle + p_2 \mid E_2\rangle \tag{6.1}$$

where $\mid p_1 \mid^2 = P_1$ and $\mid p_2 \mid^2 = P_2$.

But now let us suppose that time $t$ becomes greater than $(R_1/v)$ and that the observer does not observe a scintillation from $E_1$. Then according to the knowledge interpretation the state vector must collapse, with the result that the probabilities become $P_1 = 0$ and $P_2 = 1$, and the state vector becomes $\mid S(t)\rangle = \mid E_2\rangle$ for $t > (R_1/v)$. The interpretational problem as stated by Renninger and de Broglie is that the state vector has collapsed abruptly and non-linearly, and yet "the observer sees nothing at all on screen $E_1$, where nothing has happened". Thus, it would appear that the absence of an interaction with the measurement apparatus leading to the absence of an observation can collapse the state vector as readily as a positive and definite observation.

This *gedankenexperiment* helps us to understand the knowledge interpretation logic that led von Neumann [10] and Wigner [11] to stress the need for a conscious and intelligent observer as the triggering agent for the collapse of the state vector.

The change in "knowledge" when no scintillation is observed at $E_1$ when $t = R_1/v$ requires a deduction on the part of the observer as to what should have happened if the alpha particle had been aimed at $E_1$. It correspondingly casts some doubt on Schrödinger's principle of state distinction [7] and on Heisenberg's irreversibility criterion [12], since no state-distinguishing record is made at $t = R_1/v$ and no irreversible process is initiated. Furthermore, one could imagine a more elaborate version of this experiment with a very large number of partial spheres inside $E_2$, so complicated that no human observer could possibly keep track of all the times and expectations of flashes that would signal the occurrence or elimination of various possible outcomes. And one could speculate on how the state vector collapse might occur in that situation. We also note that Neumaier [13] posted a *gedankenexperiment* on the quantum physics arXiv that he named the "Collapse Challenge" and that is the equivalent of the Renninger *gedankenexperiment*. He points out the deficiencies of the Copenhagen Interpretation in analyzing the system and says that the decoherence interpretation "only fakes the real situation".

The Transactional Interpretation avoids the conceptual problems implicit in this experiment by eliminating any state vector collapse that occurs at some definite instant such as $t = R_1/v$. In the TI, the state vector does not change at $t = R_1/v$, as the knowledge interpretation would imply. Instead, the TI employs an atemporal four-space description implicit in the transaction model: the state vector is emitted from the source at $t = 0$ as a retarded offer wave that grows as a spherical wave front, part of which encounters $E_2$ at $t = R_2/v$ and the remainder encounters $E_1$ at $t = R_1/v$. The boundary condition of $S$ that only a single alpha particle is emitted permits one and only one transaction to occur between $S$ and $E_1$ or $E_2$. The transaction will occur with a probability proportional to the confirmation wave echoes that $S$ receives from the two possible absorbers. These echoes will be proportional to the solid angles subtended by the two possible absorbers, i.e., $\Omega_1$ and $\Omega_2$ as expected. A single transaction forms in accordance with these probabilities through the exchange of advanced and retarded waves characterizing the transition of an alpha particle from $S$ to $E_1$ or to $E_2$.

As in Sect. 6.2, we note in this experiment the Copenhagen knowledge interpretation predicts an in-flight change in the wave function moving towards $E_2$ after it reaches the radius of $E_1$ (because an observer could deduce that the particle did not hit $E_1$), while the Transactional Interpretation predicts *no such change* in the wave function. This is a significant difference in the two interpretations, but it leads to no observable consequences that could be tested.

## 6.6 Transmission of Photons Through Non-Commuting Polarizing Filters*

The behavior of quantum systems in response to measurements of non-commuting variables is often cited as one of the interpretational problems of quantum mechanics and has been used as a justification for the development of quantum logics.

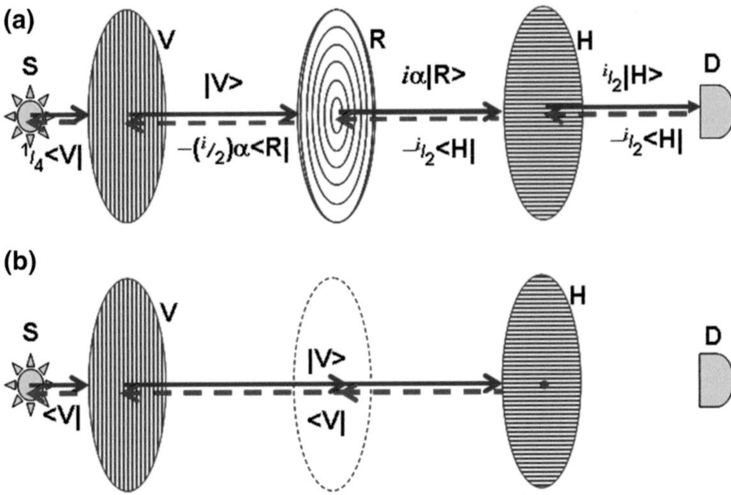


**Fig. 6.7** Schematic diagram showing **a** the passage of a single photon through successive non-commuting polarizing filters $V$, $R$, and $H$ (see text), and **b** Same diagram with filter $R$ removed. Offer waves are shown as *blue/solid* and confirmation waves as *red/dashed*

However, one can usually find excellent classical analogs of such measurements, e.g., the Fourier time-frequency complementarity of electrical pulse wave-forms (see Sect. 2.4) and the transmission of light through successive polarizing filters.

Therefore, it is instructive to consider the QM treatment of the transmission of light through polarizing filters as an illustration of the application of the TI. We will specifically select a case where the handling of complex amplitudes is required so that this aspect of the TI can be demonstrated. Figure 6.7 shows the system to be considered: A single photon of light is emitted by source $S$ and travels along an optical bench to the single-quantum detector $D$. In traversing this path, it passes through three polarizing filters, which we will call $V$, $R$, and $H$ to indicate that, respectively, they transmit with 100 % efficiency light which is in a pure state of vertical linear polarization, right circular polarization, and horizontal linear polarization, respectively, while completely absorbing light that has the orthogonal polarization. Right circular polarization means that an observer viewing an oncoming wave will see its electric vector as rotating in the counter-clockwise direction, and the trajectory of the tip of the electric field vector would trace the threads of a right-handed screw. Similarly, the observer would see electric vector of a left circularly polarized wave as rotating in the clockwise direction, and the trajectory of the tip of the electric field vector would trace the threads of a left-handed screw.

This example is chosen because the operators characterizing linear polarization eigenstates do not commute with the operators characterizing circular polarization eigenstates, and so linear and circular polarization are non-commuting variables. The two descriptions (linear vs. circular) represent two related bases. In particular, if $\mid H\rangle$, $\mid V\rangle$, $\mid R\rangle$, and $\mid L\rangle$ represent pure states, respectively, of horizontal linear, vertical linear, right circular, and left circular polarization, then they are related by the basis transform equations:

$$| R\rangle = \alpha(| H\rangle - i | V\rangle) \tag{6.2}$$
$$| L\rangle = \alpha(| H\rangle + i | V\rangle) \tag{6.3}$$
$$| H\rangle = \alpha(| R\rangle + | L\rangle) \tag{6.4}$$
$$| V\rangle = i\alpha(| R\rangle - | L\rangle) \tag{6.5}$$

where $\alpha = \sqrt{1/2}$ and $i = \sqrt{-1}$. Here, multiplication by $i$ means that the wave is shifted in phase by 90°, so that its maximum arrives $1/4$ of a period early.

The Transactional Interpretation provides the following description of the transmission of a photon from $S$ to $D$: The source $S$ produces a retarded offer wave (OW) in the form of a general state vector including all possible states of polarization. This wave then passes through filter $V$. The filter transmits only $| V\rangle$, i.e., that component of the state vector that corresponds to a state of pure vertical linear polarization (VLP). This wave then travels to filter $R$, which transmits only that component of $| V\rangle$ that is in a pure state of right circular polarization (RCP). From Eq. 6.5 this is $i\alpha | R\rangle$. This RCP wave then travels to filter $H$, which transmits only the component in a pure state of horizontal linear polarization (HLP). From Eq. 6.2, this will be $\alpha(i\alpha | H\rangle) = (i/2) | H\rangle$. This HLP offer wave then strikes $D$ as an offer to be absorbed and detected.

But according to the transaction model this is only half of the story. To confirm absorption of the incident retarded wave, the detector must produce a "time-mirrored" advanced confirmation wave (CW). This wave will be the complex conjugate of the incident offer wave and will have the form:

$$CW = OW^* = [(i/2) | H\rangle]^* = -(i/2)\langle H | \tag{6.6}$$

This advanced CW travels back along the track of the incident OW until it encounters filter $H$, where it is fully transmitted since it is already in a state of pure horizontal linear polarization.

The CW then proceeds back along the track of the OW until it reaches filter $R$, where only its RCP component is transmitted. We can use Eqs. 6.2–6.5 for changing the basis of advanced waves by taking the complex conjugates (i.e., the time reverse) of both sides of the equations to obtain a new set of transformation equations. Employing that procedure, Eq. 6.4* shows us that the transmitted CW will have the form:

$$(\alpha)[-i/2 \langle R |] = -(i/2)\alpha\langle R | \tag{6.7}$$

The CW then proceeds until it reaches filter $V$, where only its VLP component is transmitted. Eq. 6.2* shows us that the transmitted wave will be:

$$CW = i\alpha[-({}^{i}\!/_{2})\alpha\langle V \,|] = {}^{1}\!/_{4}\langle V \,| \tag{6.8}$$

Thus the source has sent out an OW of unit amplitude and has received back a CW in state ${}^{1}\!/_{4}\langle V\,|$. This then is a concrete example of the assertion that the probability of a transaction is proportional to the amplitude of the CW echo from a potential absorber and is also an illustration of the operation of the Born probability law $P = \Psi\Psi^*$. The transaction will be confirmed and the photon transmitted from $S$ to $D$ with a probability of 1/4 and will arrive at $D$ in a state of pure horizontal polarization. There will also be a probability of 3/4 that the photon will not be transmitted to $D$, but instead will be absorbed by one of the filters. These are the same transmission and absorption probabilities that are given by classical optics for the transmission of an initially horizontally polarized beam of light from $S$ to $D$.

Now consider the modification of the apparatus shown in Fig. 6.7b, in which the second filter $R$ has been removed. Now the OW is placed in a pure state of VLP by filter $V$, so that when it travels to filter $H$ it cannot be transmitted. Therefore, no OW reaches the detector $D$ and no transaction from $S$ to $D$ takes place. With filter $R$ removed, the transmission of the apparatus drops from 25 to 0 %.

The TI description of other experiments involving non-commuting variables can be constructed by employing the same procedures used above (see Sect. 6.10, for example). In each case it will be found that the probability of the quantum event under consideration is just the real and positive amplitude of the echo CW response to the OW from the emitter.

## 6.7 Wheeler's Delayed Choice Experiment (1978)*

In 1978, John A. Wheeler raised another interpretational issue [14] that is now known as Wheeler's Delayed-Choice Experiment (Fig. 6.8). Suppose that we have a Young's two slit interference apparatus as discussed in Sect. 6.1, with photons produced by a light source that illuminates two slits. The source emits one and only one photon in the general direction of the slits during the time interval chosen by the observer who is operating the apparatus. Downstream of the slits are two different measuring devices. One of these is a photographic emulsion $\sigma_1$ that, when placed in the path of the photons, will record photon's positions as they strike the emulsion, so that after many photon events, the emulsion will show a collection of spots that form a two-slit interference pattern characteristic of the photon's wavelength, momentum, and the slit separation. The other measuring device consists of a lens focusing the slit-images on photographic emulsion $\sigma_2$ at image points 1′ and 2′. A photon striking either image point tells us that the photon had passed through the slit that is imaged at that position. Therefore, detection at $\sigma_2$ constitutes a determination of the slit (1 or 2) through which the photon passed.

Such an apparatus is often used to illustrate the wave-particle duality of light. The light waves that form the interference pattern on the emulsion must have passed through both slits of the apparatus in order to interfere at the emulsion, while the

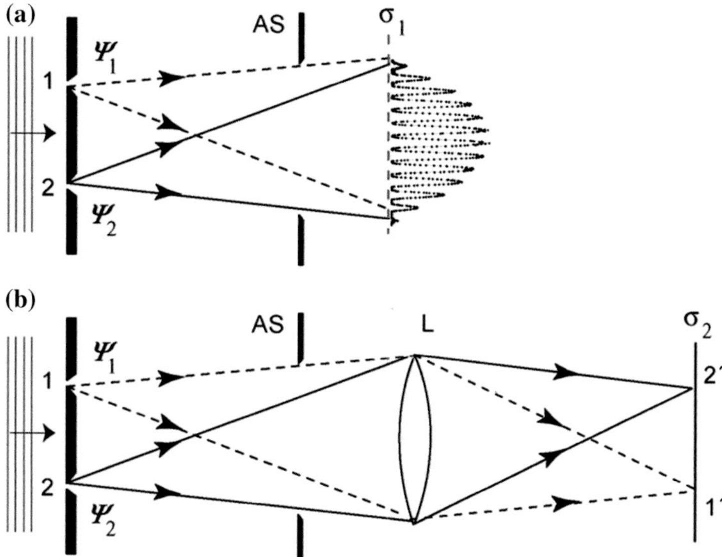


**Fig. 6.8** Wheeler's delayed choice experiment: Light from a single-photon source can either **a** produce an interference pattern on photographic emulsion $\sigma_1$ or **b** be imaged by lens $L$ to produce images of the two slits on photographic emulsion $\sigma_2$ at points $1'$ and $2'$. The experimenter waits until *after* the photon has passed through the slits to decide whether to lower photographic emulsion $\sigma_1$ so that photographic emulsion $\sigma_2$ provides which-slit information, or to leave it place so that the two-slit interference pattern characteristic of passage through both slits is observed at $\sigma_1$

photon particles that strike the photographic emulsion $\sigma_2$ can have passed through only one slit, the one imaged by the lens $L$ at image point $1'$ or $2'$. The photographic emulsion $\sigma_1$ measures momentum (and wavelength) and the photographic emulsion $\sigma_2$ measure position, i.e., conjugate variables are measured. Thus, the two experimental measurements are "complimentary" in Bohr's sense. The uncertainty principle is not violated, however, because only one of the two experiments can be performed with a given photon. But Wheeler is not done yet.

The emulsion $\sigma_1$ is mounted on a fast acting pivot mechanism, so that on command it can almost instantaneously either be raised into position to intercept the photon from the source or rapidly dropped out of the way so that the photon can proceed to $\sigma_2$. Thus when the emulsion $\sigma_1$ is up, we make an interference measurement requiring the photon to pass through both slits, and when the emulsion $\sigma_1$ is down, we make a position measurement requiring that the photon pass through only one slit.

Wheeler's innovative modification of this old *gedankenexperiment* is this: We wait until a time at which the photon has safely passed the slits but has not yet reached the emulsion apparatus $\sigma_1$. Only at that time do we decide whether to place the $\sigma_1$ emulsion up or down. The decision is made after the photon must have passed through the slit system. Therefore, the photon has already emerged from the slit system when the experimenter decides whether it should be caused to pass through one slit (emulsion down) or both slits (emulsion up). Wheeler concluded that the delayed-choice experiment illustrated his paradigm about quantum mechanics: "No phenomenon is a real phenomenon until it is an observed phenomenon."

It might be argued that there would not really be time enough for a conscious observer to make the measurement decision. However, Wheeler has pointed out that the light source might be a quasar, and the "slit system" might be a foreground galaxy that bends the light waves around both sides by gravitational lensing. Thus, there would be a time interval of millions of years for the decision to be made, during which time the light waves from the quasar were in transit from the foreground galaxy to the observer. The delayed choice experiment, since it seems to determine the path of the photon after it has passed through the slit system, has been used as an illustration of retrocausal effects in quantum processes.

The *gedankenexperiment* does not lead to any explicit contradictions, but it demonstrates some of the retrocausal implications of the standard quantum formalism. In particular, the cause (emulsion $\sigma_1$ down or up) of the change in the photon's path has come after the effect (passage through one or two slits). There have been several experimental implementations of this experiment, the most recent (2007) performed by the Aspect group in France [15]. All have shown the expected results, i.e., the predictions of standard quantum mechanics.

The Transactional Interpretation is able to give an account of the delayed choice experiment without resort to observers as collapse triggers. In the TI description the source emits a retarded OW that propagates through slits 1 and 2, producing offer waves $\psi_1$ and $\psi_2$. These reach the region of screen $\sigma_1$, where either (a) they find the screen $\sigma_1$ up and form a two-path transaction with it as illustrated in Fig. 6.8a or; (b) they find the screen $\sigma_1$ down and proceeds through lens $L$ on separate paths to screen $\sigma_2$ where they strike the screen at image points $1'$ and $2'$ and create confirmation waves that return through the lens and slits to the source. In case (b), the source receives confirmation wave echoes from two separate sites on screen $\sigma_2$ and must decide which of them to use in a one-slit competed transaction, as shown by the solid and dashed lines in Fig. 6.8b.

For case (a) in which the photon is absorbed by $\sigma_1$, the advanced confirmation wave retraces the path of the OW, traveling in the negative time direction back through both slits and back to the source. Therefore the final transaction, as shown in Fig. 6.8a, forms along the paths that pass through both slits in connecting the source with the screen $\sigma_1$. The transaction is therefore a "two-slit" quantum event. The photon can be said to have passed through *both* slits to reach the emulsion.

For case (b) the offer wave also passes through both slits on its way to $\sigma_2$. However, when the absorption takes place at one of the images (not both, because of the single quantum boundary condition), the lens focuses the confirmation wave so that it passes through only the slit imaged at the detection point. Thus the confirmation wave passes through only one slit in passing back from image to source, and the transaction which forms is characteristic of a "one-slit" quantum event. The source, receiving confirmation waves from two mutually exclusive one-slit possibilities, must choose only one of these for the formation of a transaction. The photon can be said to have passed through *only one* slit to reach $\sigma_2$.

Since in the TI description the transaction forms atemporally, the issue of *when* the observer decides which experiment to perform is not significant. The observer determined the experimental configuration and boundary conditions and the transaction formed accordingly. Further, the fact that the detection event involves a measurement (as opposed to any other interaction) is not significant and so the observer has no special role in the process. To paraphrase Wheeler's paradigm, we might say: "No offer wave is a real transaction until it is a confirmed transaction".

## 6.8 The Freedman–Clauser Experiment and the EPR Paradox (1972)*

Another quantum puzzle is the Freedman–Clauser experiment [16], previously discussed in Sect. 2.8. An atomic 2-photon cascade source produces a pair of polarization-entangled photons. If we select only entangled photons emitted back-to-back, then because of angular momentum conservation, both photons must be in the same state of circular or linear polarization. In the linear basis, their wave function should be in the Bell state of Eq. 2.1. Measurements on the photons with linear polarimeters in each arm of the experiment show that when the planes of the polarimeters are aligned, independent of the direction of alignment, the two polarimeters always measure HH or VV for the two linear polarization states, i.e., both photons are always in the same linear polarization state.

When the polarization plane of one polarimeter is rotated by an angle $\theta$ with respect to the other polarization plane, some opposite-correlation HV and VH events creep in. If $\theta$ is increased, the fraction of these events grows proportional to $1 - \cos^2(\theta)$, which for small values of $\theta$ is proportional to $\theta^2$. As discussed in Sect. 2.8, this polarization correlation behavior produces a dramatic violation of the Bell inequalities [17], which for local hidden variable alternatives to standard quantum mechanics require a growth in HV and VH events that is *linear* with $\theta$. The implication of the Bell-inequality violations is that quantum nonlocality is required to explain the observed quadratic polarization correlations.

How are the nonlocality-based polarization correlations of the Freedman–Clauser experiment possible? The Transactional Interpretation provides a clear answer, which is illustrated in Fig. 6.9. The source of the polarization-entangled photons seeks to emit the photon pair by sending out offer waves $\psi_L$ and $\psi_R$ to the left and right detectors. The detectors respond by returning confirmation waves $\psi_L*$ and $\psi_R*$ back to the source. A completed three-vertex transaction can form from these echoes, however, only if the two potential detections are compatible with the conservation of angular momentum at the source. This requirement produces the observed polarization correlations. The transaction does not depend on the separation distance of the polarimeters or on which of the polarization detection events occurs first, since

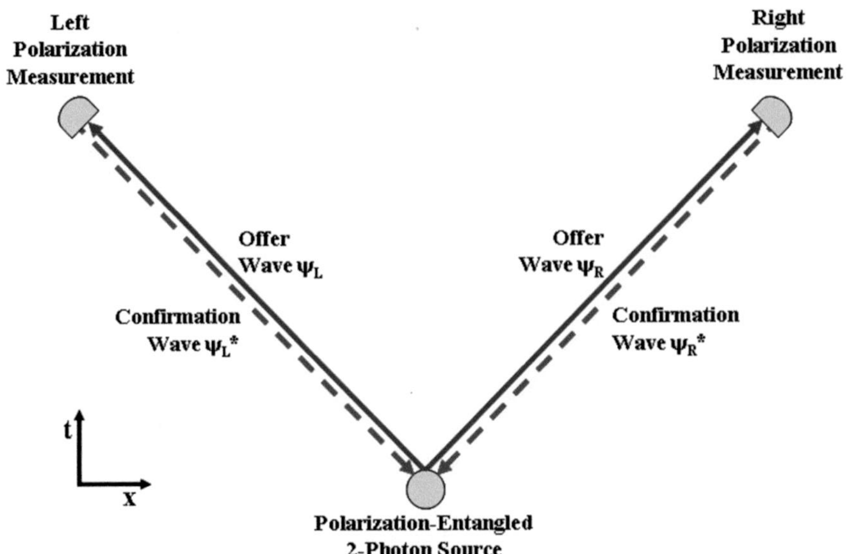


**Fig. 6.9** Space-time schematic of a nonlocal "V" transaction for visualizing the polarization-entangled Freedman–Clauser EPR experiment. Offer waves $\psi_L$ and $\psi_R$ (*blue/solid*) move from source to linear polarization detectors, and in response, confirmation waves $\psi_L*$ and $\psi_R*$ (*red/dashed*) move from detectors to source. The three-vertex transaction can form only if angular momentum is conserved by having correlated and consistent measured linear polarizations for both detected photons

the transaction formation is atemporal, and it even-handedly treats any sequence of detection events. Appendix C describes two "quantum games" that produce correlations analogous to those present in the Freedman–Clauser experiment.

## 6.9 The Hanbury Brown Twiss Effect (1956)*

The Hanbury-Brown-Twiss effect (HBT) is an example of the interference of radiation sources that are incoherent [18, 19]. It has been applied to the measurement of the diameters of nearby stars with radio interferometry and to investigation of the dimensions of the "fireball" developed in relativistic heavy ion collisions in which a large number of $\pi$-mesons (pions) are produced in each collision [20–23]. The HBT effect applies equally well to classical radio waves and to particle-like quanta such as pions.

A simplified version of a HBT interference measurement is illustrated in Fig. 6.10. Sources 1 and 2 are separated by a distance $d_{12}$. Both sources emit photons of the same energy $\hbar\omega$ but are completely incoherent. The radiation from the two sources is detected by detectors $A$ and $B$, which are separated by a distance $d_{AB}$. The line of

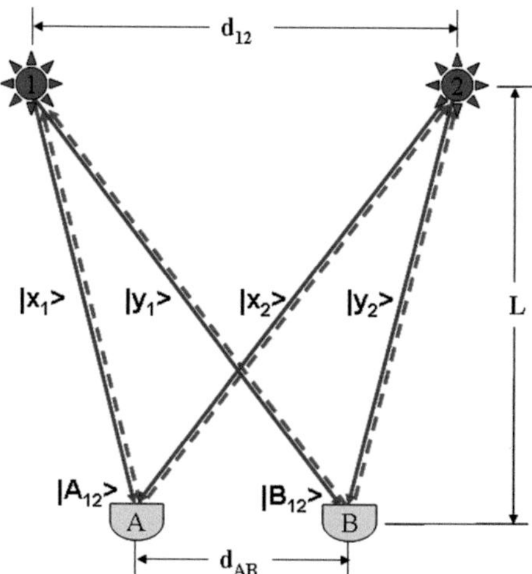


**Fig. 6.10** Schematic diagram of the Hanbury-Brown Twiss experiment demonstrating coherent interference between light or particle waves from incoherent sources. Sources 1 and 2 are separate by distance $d_{12}$, and emit offer waves $| x_{1,2}\rangle$ and $| y_{1,2}\rangle$ (*blue/solid*) of identical wavelengths $\lambda$. Detectors $A$ and $B$ located a distance $d_{AB}$ apart and a distance $L$ from the sources return confirmation waves (*red/dashed*), and 4-vertex transaction forms in which two photons are transferred and detected. A product or coincidence between detector outputs results in a composite signal exhibiting an interference effect depending on $L$, $\lambda$, $d_{12}$, and $d_{AB}$, allowing $d_{12}$ to be determined from measurements

centers of the sources is parallel to the line of centers of the detectors, and the two lines are separated by a distance $L$.

It will not be demonstrated here, but a signal that is a product of (or coincidence between) the signals received at $A$ and $B$ (indicating that photons have simultaneously triggered both detectors) reflects the coherent interference of the two sources and depends on the source separation $d_{12}$ as well as the detector separation $d_{AB}$. Measurements made at a number of values of $d_{AB}$ can therefore be used to determine $d_{12}$, in a manner analogous to moving a single detector in an interference pattern to determine the separation of a pair of coherent sources. This is the HBT intensity interference effect.

There is a lesson for applications of the Transactional Interpretation in this kind of interference phenomenon: particles like photons and pions cannot be consistently described as little blobs of mass-energy that travel from point $A$ to point $B$, as the Bohm–de Broglie interpretation would like us to believe. In the HBT effect, a whole photon is assembled at each detector out of partial-photons contributed by each of the two sources. Consider a transaction in which photons are emitted by 1 and 2 and detected by $A$ and $B$ so that their product signal exhibits HBT interference. In the TI description of such an HBT event, retarded OW's $| x_1\rangle$ and $| y_1\rangle$ are emitted by the source 1 and travel to detectors $A$ and $B$, respectively. Similarly, OW's $| x_2\rangle$ and $| y_2\rangle$ are emitted by the source 2. Detector $A$ receives a composite OW $| A_{12}\rangle$ which is a linear superposition of $|x_1\rangle$ and $|x_2\rangle$ and seeks to absorb the "offered" photon by producing advanced CW $\langle A_{12}|$, the time reverse of that superposition. Detector $B$ similarly responds to composite OW $|B_{12}\rangle$. These advanced waves then travel back to the two sources, each of which receives a different linear superposition of $\langle A_{12}|$ and $\langle B_{12}|$.

An HBT 4-vertex transaction is formed that removes one energy quantum $\hbar\omega$ from each of the two sources 1 and 2 and delivers one energy quantum $\hbar\omega$ to each of the two detectors $A$ and $B$. For many combinations of source and detector separation distances, the superimposed OW's and/or CW's are nearly equal and opposite, so that the composite wave is very weak and the transaction is very improbable. For a few ideal combinations of source and detector separation distances all of the composite waves are strong because their components coherently reinforce, and in this case the transaction is much more probable. The transaction probability depends on the separation distances in just the way predicted by quantum mechanics. Thus the HBT effect is completely explained by the Transactional Interpretation.

However, there is an interesting point here: neither of the photons detected by $A$ or $B$ can be said to have uniquely originated in one of the two sources. Each detected photon originated partly in each of the two sources. It might be said that each source produced two fractional photons and that fractions from two sources combined at a detector to make a full size photon. Particles transferred have no separate identity that is independent from the satisfaction of the quantum mechanical boundary conditions. The boundary conditions here are those imposed by the HBT geometry and detection criteria.

This two photon event may be viewed as a simple case of more general multiphoton (or multiparticle) events, which may involve many sources and many detectors. Such transactions can be viewed as assembling particles at a detector from contributions derived from an number of sources, with no one-to-one correspondence between particles emitted and particles detected except in the overall number. One way of stating this is to emphasize that the spatial localization of the emitter (or the absorber) may be very fuzzy and indefinite, so long as all boundary conditions are satisfied. Likewise the time localization of the emission event (or absorption event) can be made very indefinite by a choice of experimental conditions, e.g., very low emission probability as in the Pflegor–Mandel experiment [24].

## 6.10 The Albert–Aharonov–D'Amato Predictions (1985)

The predictions of Albert, Aharonov, and D'Amato [25] (AAD) clarify an old problem, the question of retrospective knowledge of a quantum state following successive measurements of non-commuting variables [26]. The assumption of contra-factual definiteness (CFD) plays an important role in the AAD predictions because these concern the retrospective knowledge of the observer about the outcome of experiments that might have been performed on the system in the time interval between one of the measurements and the other. We need the CFD assumptions that the various alternative possible measurements that might have been performed on the system would each have produced a definite (although unknown and possibly random) observational result and that we are permitted to discuss these results. Under the assumption of CFD, the AAD predictions provide a challenging interpretational problem.

As a simple example of the AAD predictions, consider the experiment illustrated in Fig. 6.11. A photon is emitted from source *S* and is transmitted through a filter *V* that passes only vertical linearly polarized (VLP) light. It then travels a distance *L* and is transmitted through a second filter *R* that passes only right circularly polarized (RCP) light. The photon is then detected by a quantum sensitive detector *D*, which generates an electrical signal registering the arrival of the photon. The questions that are addressed by AAD are: (1) What is the quantum state of the photon in the region L, which lies in the region between *V* and *R*, and (2) What would have been the outcome of measurements on the photon that might have been performed in that region?

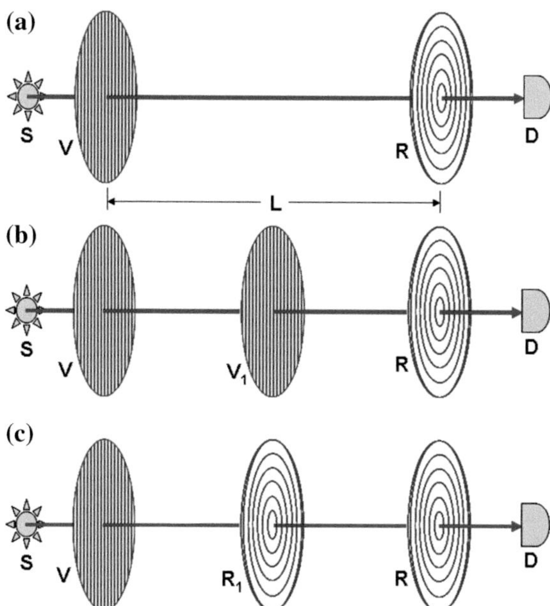


**Fig. 6.11** Schematic diagram showing the three experimental situations considered in the AAD predictions. **a** The photon emerges from the source *S*, passes through a vertical linear polarizing filter *V*, and then through a right circular polarizing filter *R* before being detected by a photomultiplier tube *D*. **b** An intermediate vertical linear polarizing filter $V_1$ is inserted. **c** An intermediate right circular polarizing filter $R_1$ is inserted. These additional measurements (**b** and **c**) are said [25] to demonstrate that the photon is simultaneously in a state of linear and circular polarization in the intermediate region

The authors of AAD use the formalism of quantum mechanics as applied to the joint probability of a series of measurements [26] to demonstrate a remarkable pair of predictions (here applied to the present example): (1) if a linear polarization measurement had been performed (Fig. 6.11b) in region L the photon would have been found to be in a VLP state, and (2) if a circular polarization measurement (Fig. 6.11c) had been performed in region $L$ the photon would have been found to be in a RCP state. In other words the intermediate measurement of polarization appears to be equally influenced by the past linear polarization measurement that was performed at $V$ and by the future circular polarization measurement that will be performed at R, in that both seem to equally prepare the system in a definite state that "forces" the outcome of the intermediate measurement.

This completely valid application of the QM formalism appears to be in at least interpretational conflict with the uncertainty principle and with complementarity, which assert that since RCP and VLP states are eigenstates of noncommuting variables, a photon cannot have been in both of these eigenstates simultaneously. The authors of AAD, on the other hand, interpret their result as indicating that "without violating the statistical predictions of quantum mechanics, it can be consistently supposed …that non-commuting observables can simultaneously be well defined" and that indeed, "given those statistical predictions, … it is inconsistent to suppose anything else". The AAD result was summarized in a popular science account as indicating that: "The measurement on Friday caused, in some sense of the word "cause", the smeared-out values of spin on Wednesday to collapse into some definite configuration. The logical puzzle about time and causality that this development engenders has not yet been fully explored."

It is therefore of considerable interest to apply the Transactional Interpretation to this interpretational puzzle, both as a means of gaining insight into the problem and as a test of the utility of the TI for resolving the interpretational paradoxes of quantum mechanics. The TI analysis of this problem follows that of Sect. 6.6, which also dealt with the transmission of a photon through polarizing filters. The three experimental configurations considered are illustrated in Fig. 6.11a–c, and Fig. 6.12a–c show diagrammatically the corresponding state vector (SV) descriptions that will be discussed. These experimental configurations must be treated as separate (but related) quantum mechanical systems and each must be analyzed separately with the TI. Let us first consider Fig. 6.11b.

The TI provides the following description of the transmission of the photon from $S$ to $D$ with an intermediate VLP measurement: The source $S$ produces a retarded offer wave (OW) in the form of a general SV including all possible states of polarization. This wave then passes through filter $V$. The filter transmit only $\mid V\rangle$, i.e., that component of the SV that corresponds to a state of pure vertical linear polarization (VLP). This wave then travels to filter $V_1$, which transmits $\mid V\rangle$ unchanged. This VLP wave then travels to filter $R$, which transmits only that component that is in a pure state of right circular polarization (RCP). From Eq. 6.5, this will be $i\alpha \mid R\rangle$, where $\alpha = {}^1\!/\!\sqrt{2}$. This RCP wave then strikes the detector $D$ and produces the advanced confirmation wave (CW) $-i\alpha\langle R \mid$, the complex conjugate of the OW at $D$, which travels back along the track of the incident OW to confirm the transaction. When the

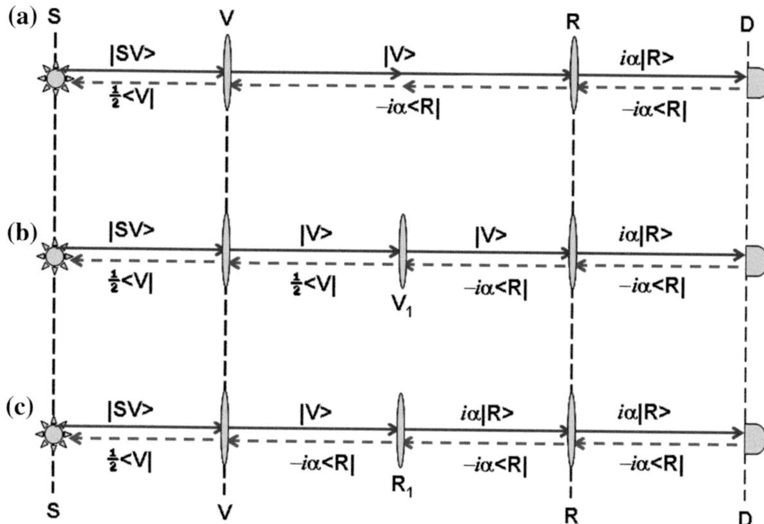


**Fig. 6.12** Schematic diagram showing the Transactional Interpretation descriptions of the three AAD experiments. Offer waves are *blue/solid* and confirmation waves are *red/dashed*. Note the differences in quantum states in the intermediate region in (**b**) and (**c**). Here $\alpha = 1/\sqrt{2}$

CW reaches $R$ it is transmitted without modification because it is already in a state of RCP. However, when it reaches $V_1$, only its VLP component is transmitted, so from Eq. 6.2* it becomes $i\alpha(-i\alpha\langle V\,|) = 1/2\,\langle V\,|$. As discussed in Sect. 6.6, we use the complex conjugates of the basis transform Eqs. 6.2–6.5 when dealing with the filtering of advanced waves. The CW retains the same form as it passes through the filter $V$ and back to the source $S$.

The description of the transmission of the photon from $S$ to $D$ with an intermediate RCP measurement illustrated in Fig. 6.12c is very similar: The source $S$ produces a retarded OW in the form of a general SV including all possible states of polarization. This wave then passes through filter $V$, which transmits only $|\,V\rangle$. This wave then travels to filter $R_1$, which transmits only that component that is in a pure state of right circular polarization (RCP). From Eq. 6.5, this will be $i\alpha\,|\,R\rangle$. This RCP wave then travels to filter $R$, which transmits $i\alpha\,|\,R\rangle$ unchanged. It reaches detector $D$ and produces the advanced CW $-i\alpha\langle R\,|$, the complex conjugate of the OW, which travels back along the track of the incident OW to confirm the transaction. When the CW reaches $R$ and $R_1$, it is transmitted without modification because it is already in a state of RCP. However, when it reaches $V$, only its VLP component is transmitted, so from Eq. 6.2* it becomes $i\alpha(-i\alpha\langle V\,|) = 1/2\,\langle V\,|$. It retains this form as it passes back to the source $S$.

In cases (b) and (c) the insertion of the intermediate polarizing filter does not alter the statistical aspects of the measurement from that of case (a) where there is no intermediate measurement, and so the three cases are equivalent in the observational sense. However, the TI gives us the opportunity to examine the intermediate quantum states in each case, and when this is done we find that the transaction that is confirmed is quite different in each of the three cases. This is illustrated in Fig. 6.12. In case (a) where there is no intermediate measurement the state in the intermediate region between $V$ and $R$ is in an indeterminate quantum state, in that the OW is $| V\rangle$ while the CW is $-i\alpha\langle R |$. This is also the case for the region between $V_1$ and $R$ for case (b) and for the region between $V$ and $R_1$ for case (c). However, we see that for case (b) the CW in the region between $V$ and $V_1$ is in a state of pure VLP, while for case (c) the CW between $R_1$ and $R$ is in a state of pure RCP.

The TI resolution of the riddle posed by the AAD predictions is that the uncertainty principle is not compromised, nor can non-commuting observables simultaneously be well defined, as the AAD authors have suggested. However, as was suggested above in another context, the circular polarization measurement that occurs later at $R$ does cause, in some sense of the word cause, the smeared-out values of circular polarization between $R$ and $V$ to earlier "collapse into some definite configuration". The transactions that form in the three cases are not identical, even though they lead to the same observables, because each transaction is a separate self-consistent solution to the wave equation. Each satisfies a different set of boundary conditions. The insertion of the intermediate filter, while not altering the statistics of the measurement, brings into being a different transaction that has different characteristic eigenstates in the intermediate region between $V$ and $R$. Thus, the two predictions of the AAD calculation concern intrinsically different quantum systems and cannot be construed as implying the presence "simultaneously" of the eigenstates of no-commuting variables, as was incorrectly asserted.

## 6.11 The Quantum Eraser (1995)*

A more elaborate delayed-choice variation is the quantum eraser experiment, a high-tech descendant of Wheeler's delayed choice concept. The experiment used a new (in 1995) trick for making "entangled" quantum states. If ultraviolet light from a 351 nanometer (nm) argon-ion laser passes through a $LiIO_3$ crystal, non-linear effects in the crystal can "split" the laser photon into two longer wavelength photons at 633 nm and 789 nm in a process called "down-conversion". The energies of these two "daughter" photons add up to the energy of their pump-photon parent, as do their vector momenta, and they are connected non-locally because they constitute a single "entangled" quantum state. They are required to be in correlated states of polarization, and under the conditions of this down-conversion they will be vertically polarized. As in other EPR experiments, a measurement performed on one of these photons affects the outcome of measurements performed on the other.

In a version of the experiment performed by Anton Zeilinger's group in Innsbruck, Austria, [27] the laser beam is reflected so that it makes two passes through the non-linear crystal, so that an entangled photon pair may be produced in either the first or the second pass through the non-linear crystal. As shown in Fig. 6.13, the experiment has the configuration of a six-pointed star formed of three beam paths intersecting

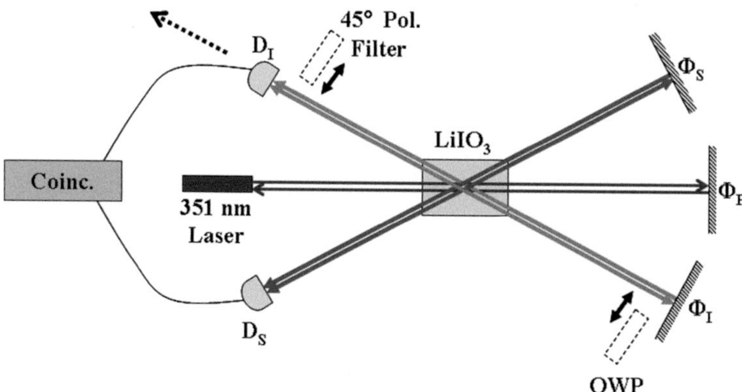


**Fig. 6.13** Schematic diagram of the quantum eraser experiment. A $LiIO_3$ nonlinear crystal is pumped by a 351 nm laser beam (*violet*) and produces by down-conversion vertically polarized 633 nm (*orange*) and 789 nm (*red*) photons that can be made in either pump-photon pass through the crystal. A quarter-wave plate (QWP) and 45° polarizing filter may be inserted in the $I$ path and the path to $D_I$ may be lengthened (see text)

at a point inside the crystal. The laser beam first passes through the crystal moving horizontally downstream, is reflected by a downstream mirror $\Phi_P$, and then passes through the crystal again moving horizontally upstream. Along the two diagonal branches downstream of the laser the two down-converted photons made in the first laser-pass travel to mirrors $\Phi_S$ and $\Phi_I$ ($S$ for signal and $I$ for idler), where they are reflected back to their production point and travel past it to upstream detectors $D_S$ and $D_I$. The laser beam, in making its second pass through the crystal has a second chance to make a pair of down-converted photons. If these are produced, they travel directly to the upstream detectors along the two upstream diagonal branches.

The net result is that a photon arriving in coincidence at the two upstream detectors may have been produced in either the first laser pass through the crystal and then reflected to the detector, or in the second pass and traveled directly to the detector. There is no way of determining which "history" (direct vs. reflected) happened, so the states are superimposed. Therefore, the quantum wave functions describing these two possible production histories must interfere. The interference may be constructive or destructive, depending on the interference phase determined by the downstream path lengths (all about 13 cm) to the three mirrors of the system. Changing the path length to one of the mirrors (for example, by moving the laser-beam reflector $\Phi_P$) is observed to produce a succession of interference maxima and minima in the two detectors.

This experimental setup is governed by the same physics as the delayed-choice experiment of Sect. 6.7, but, because there are two coincident photons and well separated paths for the two possible histories, it is easier to play quantum tricks with the system. Initially, all polarizations are vertical. Now the experiment is modified to remove the quantum interference by placing distinguishing polarization labels on the two possible photon histories (direct vs. reflected). A transparent optical element called a "quarter-wave plate" (QWP) is placed in front of the photon reflection mirror $\Phi_I$. The QWP is set to rotate the polarization state of the reflected photons from vertical to horizontal polarization as they pass twice through it. This polarization modification allows the reflected and direct "histories" to be quantum-distinguishable, because one of the reflected photons is horizontally polarized while the direct photons are vertically polarized. The two superimposed quantum states are now distinguishable (even if no polarization measurement is actually made), and the interference pattern is eliminated, both in the *I* arm of the experiment in which the QWP is placed and also in the other *S* arm, where no modification was made.

Finally, the "quantum eraser" is brought into use. Any vertically or horizontally polarized light beam can be separated into a light component polarized 45° to the left of vertical and a light component polarized 45° to the right of vertical. Therefore, for the photons with the QWP in front of their mirror, placing just in front of their detector a filter that passes only light polarized 45° to the left of vertical "erases" the label that had distinguished the two histories by making the polarizations of the two waves reaching detector $D_I$ the same. When this is done, it is found that interference is restored.

Further, the paths to the two detectors can have different lengths, with the path through the 45° filter to $D_I$ made much longer than the path to detector $D_S$. This has the effect of erasing the path-distinguishing label on the *I* photon *after* the *S* photon had already been detected. This modification is observed to have no effect on the interference. The *post-facto* erasure still restores interference. The path label can be erased retroactively and has the same effect (retroactive or not) on the quantum interference of the waves. Effectively, the quantum eraser has erased the past!

The Transactional Interpretation can easily explain the curious retroactive erasure of "which-way" information. When which-way information is present, separate transactions must form for each of the paths, and no interference is observed. When the which-way information is erased, the overall transaction that forms involves both paths, and interference is observed. Modifying the polarizations causes a different type of transaction formation, resulting in different observations. The retroactive erasure of the which-way information is irrelevant, because the transaction forms atemporally, connecting the source and detectors in one or two advanced-retarded TI handshakes across space-time.

## 6.12 Interaction-Free Measurements (1993)*

In 1993, Elitzur and Vaidmann [28] (EV) showed a surprised physics community that quantum mechanics permits the non-classical use of light to examine an object without a single photon of the light actually interacting with the object. The EV experiment requires only the *possibility* of an interaction.

In their paper [28] Elitzur and Vaidmann discuss their scenario in terms of the standard Copenhagen Interpretation of quantum mechanics, in which the interaction-free

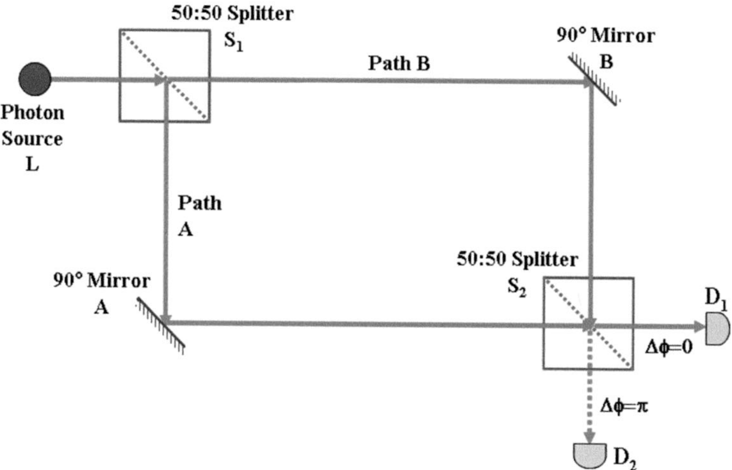


**Fig. 6.14** Mach Zehnder interferometer with both beam paths open. All photons go to $D_1$ because of destructive interference at $D_2$

result is rather mysterious, particularly since the measurement produces "knowledge" that is not available classically. They also considered their scenario in terms of the Everett–Wheeler or "many-worlds" interpretation of quantum mechanics [29, 30]. Considering the latter, they suggest that the information indicating the presence of the opaque object can be considered to have come from an interaction that had occurred in a separate Everett–Wheeler universe and was transferred to our universe through the absence of interference. Here we will examine the same scenario in terms of the Transactional Interpretation and will provide a more plausible account of the physical processes that underlie interaction-free measurements.

The basic apparatus used by EV is a Mach–Zender interferometer, as shown in Fig. 6.14. Light from a light source $L$ goes to a 50:50 % beam splitter $S_1$ that divides incoming light into two possible paths or beams. These beams are deflected by 90° by mirrors $A$ and $B$, so that they meet at a second beam splitter $S_2$, which recombines them by another reflection or transmission. The combined beams from $S_2$ then go to the photon detectors $D_1$ and $D_2$.

The Mach–Zehnder interferometer has the characteristic that, if the paths $A$ and $B$ have precisely the same path lengths, the superimposed waves from the two paths are in phase at $D_1$ ($\Delta\phi = 0$) and out of phase at $D_2$ ($\Delta\phi = \pi$). This is because with beam splitters, an emerging wave reflected at 90° is always 90° out of phase with the incident and transmitted waves [31]. The result is that all photons from light source $L$ will go to detector $D_1$ and none will go to detector $D_2$.

Now, as shown in Fig. 6.15 we place an opaque object (*Obj*) on path $A$. It will block light waves along the lower path after reflection from mirror $A$, insuring that all of the light arriving at beam splitter $S_2$ has traveled there via path $B$. In this case

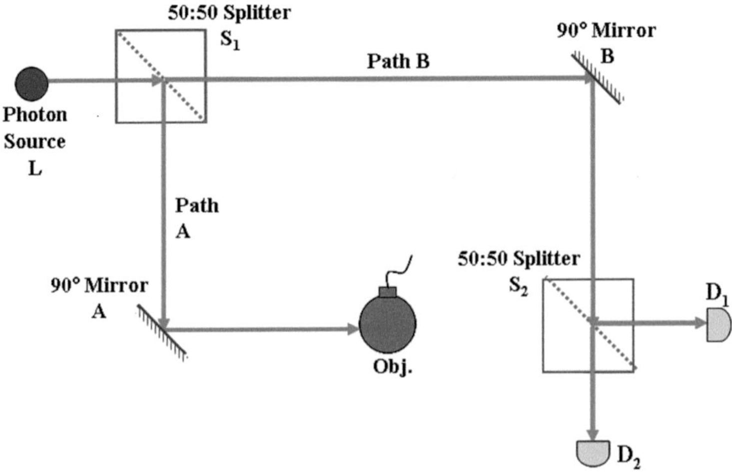


**Fig. 6.15** Mach Zehnder interferometer with one beam path blocked. Half of the photons are absorbed by the blocking object, 25 % go to $D_1$, and 25 % go to $D_2$

there is no interference, and beam splitter $S_2$ sends equal components of the incident wave to the two detectors.

Now suppose that we arrange for the light source $L$ to emit only one photon within a given time period. Then, if we do the measurement with no opaque object on path $A$, we should detect the photon at $D_1$ 100 % of the time. If we perform the same measurement with the opaque object *Obj* blocking path $A$, we should detect a photon at $D_1$ 25 % of the time, a photon at $D_2$ 25 % of the time, and should detect no photon at all 50 % of the time (because it was removed by *Obj* in path $A$). In other words, the detection of a photon at $D_2$ guarantees that an opaque object is blocking path $A$, although no photon had actually interacted with object *Obj*. This is the essence of the Elitzur and Vaidmann interaction-free measurement.

Note that if a photon is detected at detector $D_1$, the issue of whether an object blocks path $A$ is unresolved. However, in that case another photon can be sent into the system, and this can be repeated until either a photon is detected at $D_2$ or absorbed by *Obj*. The net result of such a recursive procedure is that 66 % of the time a photon will strike the object, resulting in no detection signal, while 33 % of the time a photon will be detected at $D_2$, indicating without interaction that an object blocks the $A$ path. Thus, the EV procedure has an efficiency for non-interactive detection of 33 %.

As before, in analyzing interaction-free measurements with the Transactional Interpretation, we will explicitly indicate offer waves $\psi$ by a specification of the path in a Dirac *ket* state vector $\psi = \mid path\rangle$, and we will underline the symbols for optical elements at which a reflection has occurred. Confirmation waves $\psi^*$ will similarly be indicated by a Dirac *bra* state vector $\psi^* = \langle path \mid$, and will indicate the path considered by listing the elements in the time-reversed path with reflections underlined.

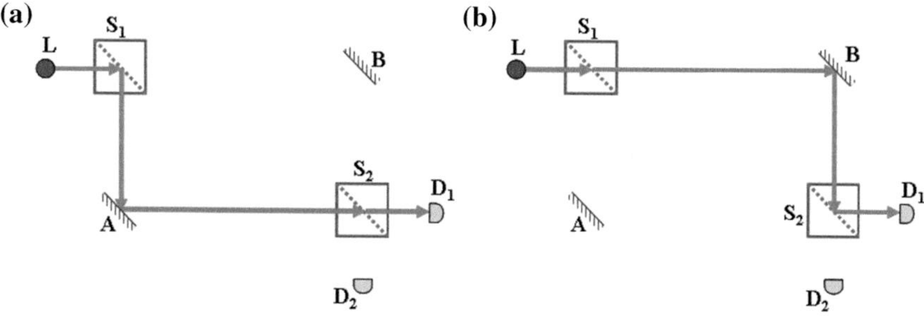


**Fig. 6.16** Offer waves **a** $| L\text{-}\underline{S_1}\text{-}\underline{A}\text{-}S_2\text{-}D_1\rangle$ and **b** $| L\text{-}S_1\text{-}\underline{B}\text{-}\underline{S_2}\text{-}D_1\rangle$

Consider first the situation in which no object is present in path $A$ as shown in Fig. 6.16. The offer waves from $L$ to detector $D_1$ are $| L\text{-}\underline{S_1}\text{-}\underline{A}\text{-}S_2\text{-}D_1\rangle$ and $| L\text{-}S_1\text{-}\underline{B}\text{-}\underline{S_2}\text{-}D_1\rangle$. They arrive at detector $D_1$ in phase because the offer waves on both paths have been transmitted once and reflected twice. The offer wave from $L$ initially has unit amplitude, but the splits at $1/\sqrt{2}$ each reduce the wave amplitude by $1/\sqrt{2}$ so that each wave, having been split twice, has an amplitude of $1/2$ as it reaches detector $D_1$. Therefore, the two offer waves of equal amplitude and phase interfere constructively, reinforce, and produce a confirmation wave that is initially of unit amplitude.

Similarly, the offer waves from $L$ to detector $D_2$ are $| L\text{-}\underline{S_1}\text{-}\underline{A}\text{-}\underline{S_2}\text{-}D_2\rangle$ and $| L\text{-}S_1\text{-}\underline{B}\text{-}S_2\text{-}D_2\rangle$. They arrive at detector $D_2$ 180° out of phase, because the offer wave on path $A$ has been reflected three times while the offer wave on path $B$ has been transmitted twice and reflected once. Therefore, the two waves with amplitudes $\pm i/2$ interfere destructively, cancel at detector $D_2$, and produce no confirmation wave.

The confirmation waves from detector $D_1$ to $L$ are $\langle D_1\text{-}S_2\text{-}\underline{A}\text{-}\underline{S_1}\text{-}L \,|$ and $\langle D_1\text{-}\underline{S_2}\text{-}\underline{B}\text{-}S_1\text{-}L \,|$. They arrive back at the source $L$ in phase because, as in the previous case, the confirmation waves on both paths have been transmitted once and reflected twice. As before the splits at $S_1$ and $S_2$ each reduce the wave amplitude by $1/\sqrt{2}$, so that each confirmation wave has an amplitude of $1/2$ as it reaches source $L$. Therefore, the two offer waves interfere constructively, reinforce and have unit amplitude. Since the source $L$ receives a unit amplitude confirmation wave from detector $D_1$ and no confirmation wave from detector $D_2$, the transaction forms along the path from $L$ to $D_1$ via $A$ and $B$. The result of the transaction is that a photon is always transferred from the source $L$ to detector $D_1$ and that no photons can be transferred to $D_2$. Note that the transaction forms along *both* paths from $L$ to $D_1$. This is a transactional account of the operation of the Mach–Zender interferometer.

Now let us consider the situation when the object blocks path $A$ as shown in Fig. 6.17. The offer wave on path $A$ is $| L\text{-}\underline{S_1}\text{-}\underline{A}\text{-}Obj\rangle$. As before an offer wave on path $B$ is $| L\text{-}S_1\text{-}\underline{B}\text{-}\underline{S_2}\text{-}D_1\rangle$, and it travels from $L$ to detector $D_1$. The wave on path $B$ also splits at $S_2$ to form offer wave $| L\text{-}S_1\text{-}\underline{B}\text{-}S_2\text{-}D_2\rangle$, which arrives at detector $D_2$. The splits at $S_1$ and $S_2$ each reduce the wave amplitude by $1/\sqrt{2}$, so that the offer wave at each detector, having been split twice, has an amplitude of $1/2$. However,

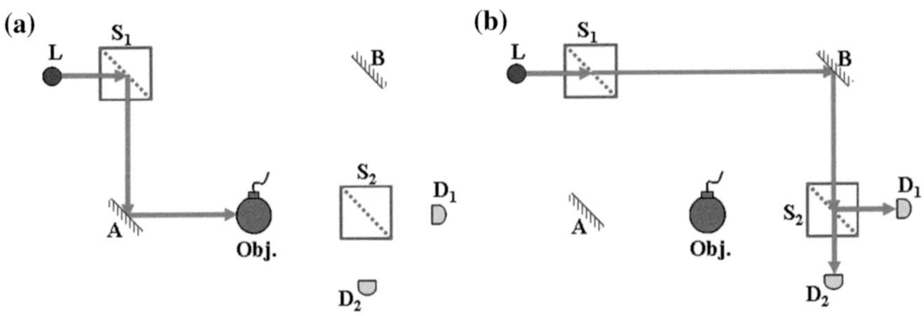


**Fig. 6.17** **a** Offer waves $| L\text{-}\underline{S_1}\text{-}\underline{A}\text{-}Obj\rangle$ and **b** $| L\text{-}S_1\text{-}\underline{B}\text{-}\underline{S_2}\text{-}D_1\rangle + | L\text{-}S_1\text{-}\underline{B}\text{-}S_2\text{-}D_2\rangle$

the offer wave $| L\text{-}\underline{S_1}\text{-}\underline{A}\text{-}Obj\rangle$ to the object in path $A$, having been split only once, is stronger and has amplitude of $^1/_{\sqrt{2}}$.

In this situation, the source $L$ will receive confirmation waves from both detectors and also from the object. These, respectively, will be confirmation waves $\langle D_1\text{-}\underline{S_2}\text{-}\underline{B}\text{-}S_1\text{-}L\,|$, $\langle D_2\text{-}S_2\text{-}\underline{B}\text{-}S_1\text{-}L\,|$ and $\langle Obj\text{-}\underline{A}\text{-}\underline{S_1}\text{-}L\,|$. The first two confirmation waves started from their detectors with amplitudes of $^1/_2$ (the final amplitude of their respective offer waves) and have subsequently been split twice. Therefore, they arrive at source $L$ with amplitudes of $^1/_4$. On the other hand, the confirmation wave from the object initially has amplitude $^1/_{\sqrt{2}}$, and it has been split only once, so it arrives at the source with amplitude $^1/_2$.

The source $L$ has one photon to emit and three confirmations to choose from, with round-trip amplitudes $(\psi\psi^*)$ of $^1/_4$, to $D_1$ $^1/_4$ to $D_2$, and $^1/_2$ to object $Obj$. In keeping with the probability assumption of the Transactional Interpretation and Born's probability law, it will choose with a probability proportional to these amplitudes. Therefore, the emitted photon goes to $D_1$ 25 % of the time, to $D_2$ 25 % of the time, and to object $Obj$ in path $A$ 50 % of the time. As we have seen above, the presence of the object in path $A$ modifies the detection probabilities so that detector $D_2$ will receive $^1/_4$ of the emitted photons, rather than none of them, as it would do if the object were absent.

How can the transfer of non-classical knowledge be understood in terms of the transactional account of the process? In the case where there is an object in the $A$ path, it is probed both by the offer wave from $L$ and by the aborted confirmation waves from $D_1$ and $D_2$. The latter are 180° out of phase and cancel. When we detect a photon at $D_2$, (i.e., when a transaction forms between $L$ and $D_2$), the object has not interacted with a photon (i.e., a transaction has not formed between $L$ and the object $Obj$). However, it has been *probed* by an offer wave from the source, which “feels” its presence and modifies the interference balance at the detectors, providing non-classical information. Thus, the Transactional Interpretation gives a simple explanation of the mystery of interaction-free measurements.

## 6.13 The Quantum Zeno Effect (1998)*

P.G. Kwiat et al. [32], a collaboration based at Los Alamos National Laboratory and the University of Innsbruck, have demonstrated both theoretically and experimentally that the efficiency of an interaction-free measurement can be increased from 33 % in the EV scheme to a value that is significantly larger. In fact, the efficiency can be made to approach 100 %, depending on how many times N it is possible to cycle the incident photon through the measurement apparatus. Their scheme is shown in Fig. 6.18.

Here a light source $L$ supplies photons that are horizontally ($H$) polarized. These are injected ($In$) into an optical "racetrack" that is capable of cycling a photon around in a closed rectangular loop $N$ times before extracting it ($Out$) to an analyzing system. After injection, the photon passes through an optical polarization rotator element ($R$) that changes its direction of linear polarization by an angle $\theta = \pi/2N$. Note that if $N$ is large, this rotation is small.

The photon then travels to a polarizing beam splitter ($S_1$) that transmit horizontally polarized ($H$) light and reflects vertically polarized ($V$) light. The object ($Obj$) to be measured may (or may not) be placed in the $V$ beam path. Downstream of the object position, the $H$ and $V$ photon components enter a second polarizing beam splitter ($S_2$) that recombines them into a single beam. The recombined photon then cycles back through the apparatus. After $N$ cycles, the photon is extracted and sent to a third polarizing beam splitter ($S_3$) that, depending on the photon's polarization, routes it to a pair of photon detectors $D_H$ and $D_V$. This detection is, in effect, a measurement of whether the photon's final polarization is horizontal or vertical.

If no object is in the $V$ path, the polarization split and recombination has no net effect. The polarization rotator rotates the plane of polarization $N$ times, each time by an angle of $\pi/2N$. The cumulative rotation is therefore a rotation of $\pi/2$. Therefore,

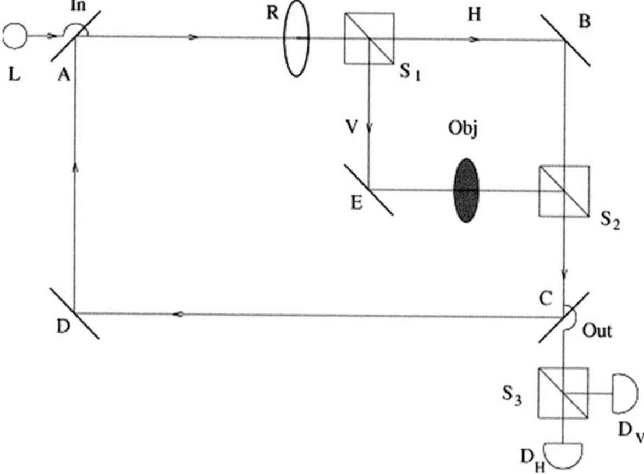


**Fig. 6.18** Quantum Zeno arrangement for high efficiency interaction-free measurements

a photon that was initially polarized horizontally ($H$) will emerge from the apparatus with vertical ($V$) polarization and will be detected by photon detector $D_V$ only.

On the other hand, if an object is placed in the $V$ path, the $H$ and $V$ beams are not recombined, so the split at the first polarization beam splitter ($S_1$) is in effect a polarization measurement. From Malus' Law, there is a probability $P_E = \cos^2(\pi/2N)$ that the photon will survive each such horizontal polarization measurement and emerge in a pure state of horizontal ($H$) polarization. After each cycle in which the photon survives, it is reset to its initial state of horizontal ($H$) polarization, so that when it is extracted after $N$ cycles it will be detected by photon detector $D_H$ only. In each cycle, there is a small probability $(1 - P_E)$ that the photon will be projected into a state of pure vertical ($V$) polarization, will travel on the $V$ path, will interact with the object, and will be removed from the process.

In summary, if the object is not present, the emerging photon will be detected by the $D_V$ detector 100 % of the time. If the object is present, the emerging photon will be detected by the $D_H$ detector with a probability $P_D = P_E^N = \cos^{2N}(\pi/2N)$, and the photon will interact with the object and be removed with a probability $P_R = 1 - P_E^N = 1 - \cos^{2N}(\pi/2N)$. We note that when $N$ is large, $P_D \approx 1 - (\pi/2)^2/N$ and $P_R \approx (\pi/2)^2/N$. Therefore, the probability of removal decreases as $1/N$ and goes to zero as $N$ goes to infinity. Therefore, the procedure greatly improves the efficiency of interaction-free measurements. For example, when the number of passes $N$ is equal to 5 the measurement is 60 % efficient. With $N = 10$ it is 78 % efficient, and with $N = 20$ it is 88 % efficient. Figure 6.19 shows an unfolding of the quantum-Zeno interaction-free measurement, for the case where no object is placed in the $V$ beam.

The recycled path is represented as a linear sequence of incremental rotations, beam splittings, and beam recombinations. It should be clear from the diagram that the successive splittings and recombinations have no net effect. On the other hand, the $N$ successive rotations have the cumulative effect of a $\pi/2$ rotation that converts the initial horizontally polarized photons into vertically polarized photons by the time they reach the final beam splitter and the detector $D_V$.

From the point of view of the Transactional Interpretation, the initial offer wave leaves the light source $L$ and is then successively rotated, split, and recombined. These operations do not reduce the amplitude, so the offer wave reaches detector $D_V$ at full strength. The confirmation wave from $D_V$ travels back along the same path and arrives back at $L$ at full strength, thereby completing the transaction.

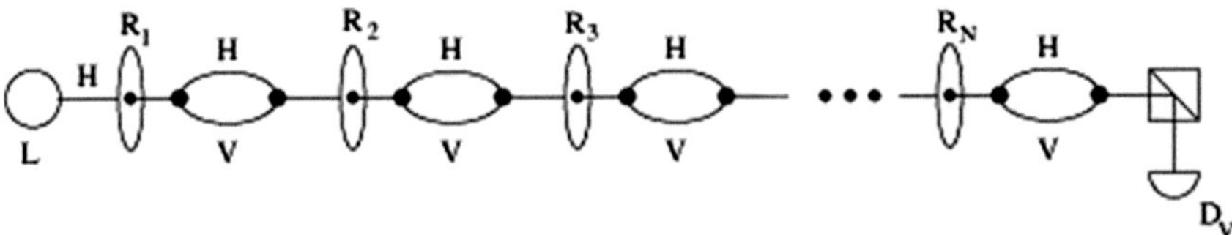


**Fig. 6.19** Unfolding of the Quantum Zeno measurement (no object) for high efficiency interaction-free measurements

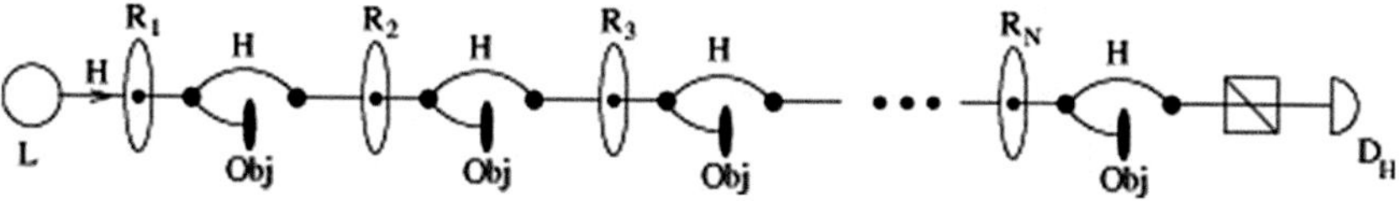


**Fig. 6.20** Unfolding of the Quantum Zeno measurement with an object present

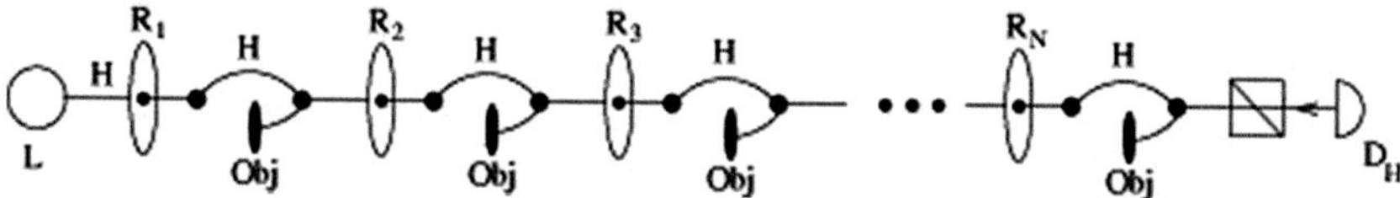


**Fig. 6.21** Paths of detector confirmation waves in the unfolded Quantum Zeno measurement with an object present

The situation when an absorber is present and there is no interaction is shown in Fig. 6.20.

Now the object *Obj* blocks the path $V$ of the vertically polarized beam after the splitter, so only the photons on the $H$ path can reach the detector system. The net effect of this is that after each incremental rotation, the beam is reset to the $H$ state and passes straight through the final splitter without deflection to reach detector $D_H$.

From the point of view of the Transactional Interpretation, the initial offer wave leaves the light source $L$ and at each rotation and splitting the intensity of the offer wave that will reach detector $D_H$ is reduced by $\cos(\pi/2N)$, so that the net intensity at the detector is $\cos^N(\pi/2N)$. At the $m$th split (for m = 1 to N), an offer wave of intensity $\cos^{m-1}(\pi/2N)\sin(\pi/2N)$ travels to the object *Obj* and may interact with it. The confirmation wave from each of these potential interactions will travel back to the light source $L$ with the same reduction factor, so that the net probability of an interaction following the $m$th split is $\cos^{2(m-1)}(\pi/2N)\sin^2(\pi/2N)$.

The path of confirmation waves from the detector $D_H$ is shown in Fig. 6.21. The confirmation wave leaves detector $D_H$ with an amplitude of $\cos^N(\pi/2N)$, the final amplitude of the offer wave. As the confirmation wave travels back to the light source $L$, at each of the $N$ splits it is reduced in intensity by a factor of $\cos(\pi/2N)$. Thus its net intensity at $L$ will be $\cos^{2N}(\pi/2N)$, which is just the probability that the detection event will occur. At each split, there is a component of the confirmation wave that takes the lower path in Fig. 6.21 and ends at object *Obj*. However, these components cannot form a transaction, since they cannot connect back to the light source $L$.

As before, the object *Obj* in the $V$ path is probed both by the offer wave from $L$ and by the aborted confirmation wave from $D_H$. When we detect a photon at $D_H$, (i.e., when a transaction forms between $L$ and $D_H$), the object *Obj* has not interacted with a photon (i.e., a transaction has not formed between $L$ and *Obj*), but the object has been probed repeatedly by weak offer and confirmation waves from both sides. As the number of passes $N$ is increased and the efficiency of the measurement approaches 100 %, the amplitudes of these probe waves grows weaker as their number increases. It also becomes clear why, even when the object does not interact with a photon, the *possibility* of interactions is required. If the interaction probability were zero, the offer and confirmation waves would not be blocked by the interposed object and the measurement would not have been possible.

## 6.14 Maudlin's *Gedankenexperiment* (1996)

In a book publication of his PhD Thesis in 1996, Tim Maudlin [33] constructed a *gedankenexperiment* that he claimed cast doubt on the validity of the Transactional Interpretation.[1]

Figure 6.22a shows Maudlin's *gedankenexperiment*. A particle source $S$ is configured to emit a single slow particle ($v \ll c$) that has a 50 % chance of being emitted in the direction of particle detector $A$ on the right and a 50 % chance of being emitted in the direction of particle detector $B$ on the left. However, particle detector $B$ is initially positioned behind $A$ on the right, and only if $A$ *does not detect the particle* is $B$ moved to its final position on the left, where the emitted particle will be detected. Maudlin claims that there cannot be the a transactional "echo" competition between the two possible outcomes because the second outcome (detection at $B$) is causally connected to the first outcome (non-detection at $A$). He also claims that the process is deterministic rather than stochastic and that after a non-detection at $A$, the wave function for detection at $B$ must have an increased amplitude, not provided by the Transactional Interpretation, because it will then have a 100 % probability of being detected. Lewis [35] has subsequently made the same latter claim.

We will dispose of Maudlin's last point first. For the purposes of evaluating probabilities including a non-detection event, his experiment is the same as that of Renninger, which is discussed in Sect. 6.6. Renninger's *gedenkenexperiment* was also discussed in some detail in Sect. 4.1 of the 1986 TI paper [36], the target of Maudlin's critique. The claim by Maudlin and Lewis that the left-going wave function should change at the instant when detector $A$ can, but does not, detect a particle. This expectation is an unverifiable prediction of Heisenberg's knowledge interpretation. It is not suggested by either the Transactional Interpretation or the standard quantum formalism. In the TI description, the left-going wave function does not magically change in mid-flight, nor does it need to. In the TI the amplitude of the left-going wave and its echo remain unchanged after a right-going particle should have reached detector $A$, and they correctly predict that 50 % of the particles will be detected by detector $B$

[1] Ruth Kastner [34] states that "Maudlin's Challenge" was taken as fatal to the Transactional Interpretation by philosophers of science for over a decade. My initial reaction to his paradox, after obtaining Maudlin's book by inter-library loan, was that before the assertions could be taken seriously, he needed to provide a mathematical description of his *gedankenexperiment* using the formalism of quantum mechanics, and then to use that formalism with the Born rule to calculate probabilities, etc., since it is the formalism, not the interpretation, that should be used to make predictions. However, at a meeting is Sydney in 2005, Kastner convinced me that Maudlin did have a point worth considering (and discussed here). This resulted in my realization that the concept of *hierarchy* needed to be added to the transaction model (see Sect. 5.5) to deal with the problem that he had raised.

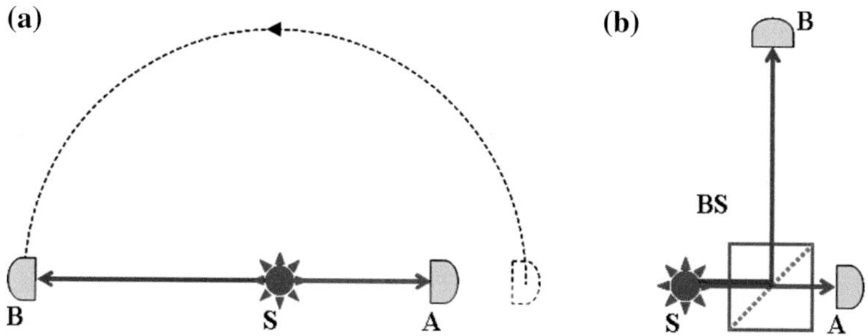


**Fig. 6.22** **a** Maudlin's *gedankenexperiment*: Slow particle source $S$ emits one particle that either goes to detector $A$ on the right (*red*) or to detector $B$ on the left (*blue*), with a 50 % probability of each. However, detector $B$ is initially positioned behind $A$ and is only moved to it final position on the left if $A$ does not detect a particle. **b** Related experiment: Single photon source $S$ sends one photon through a 50:50 splitter $BS$ to detectors $A$ and $B$

(See Sect. 6.6 for further discussion of this point.). Maudlin and Lewis have provided another example of the dangers of swallowing whole the knowledge interpretation, with its disappearing waves.

Now let's consider the Maudlin experiment and its implications. Maudlin's specifications for the source seem innocent enough, but in the real world they present a problem. Real sources in the quantum optics laboratory do not emit single particles "on command", like a gun when the trigger is pulled; they emit at a rate that is adjusted so that, in a given time interval, the probability of emitting one particle is greater than that of emitting zero or two particles, in keeping with the time-energy constraints of the uncertainty principle. Therefore, there is always a non-zero probability that no particles at all will be emitted by the source, and that possibility cannot be switched off. Thus, in the real world there are three possible outcomes for Maudlin's setup: detection at $A$, detection at $B$, and no detection at all.

How is Maudlin's experiment described with the quantum formalism? The state vector can be written as:

$$\mid S(t)\rangle = \frac{1}{\sqrt{2}}(\mid 1\rangle_A \mid 0\rangle_B + e^{i\phi} \mid 0\rangle_A \mid 1\rangle_B) \tag{6.9}$$

where $\phi$ is a phase that depends on relative path length, $\mid 1\rangle_A$ indicates detection of one photon at $A$, $\mid 1\rangle_B$ indicates detection at $B$, $\mid 0\rangle_A$ indicates no detection at $A$, and $\mid 0\rangle_B$ indicates no detection at $B$. This is the same wave function that could be used to describe the related experiment shown in Fig. 6.22b. In both experiments the external world can be undergoing irreversible changes during the time interval between the potential $A$ and $B$ detections, but only in Fig. 6.22a do these changes affect the $B$ detection configuration. In both experiments, applying the Born probability rule to Eq. 6.9 leads to 50:50 detection probabilities, in contradition to Maudlin's assertion.

Interestingly, Eq. 6.9 is a Bell-state wave function (see Sect. 2.8 and Eq. 2.2) describing the entanglement of detection at one detector with the non-detection at the other detector. The difference between the two experiments, despite the fact that they share the same wave function, is that in Fig. 6.22a a causal connection has been implemented between non-detection at $A$ and the position of $B$, while in Fig. 6.22b there is no such connection. The point of this comparison is that the QM formalism is indifferent to whether there is a causal connection or not.

However, Maudin has a valid point. There would seem to be a problem in the situation in which some early possible outcomes of a transaction can change the physical configuration of later possible outcomes. This problem is solved by the assumption of *hierarchy*. The hierarchy part of the 3D transaction model discussed in Sect. 5.5 asserts that in selecting a transaction from among the advanced-retarded wave echos, the emitter makes an ordered separate decision to select or not select each particular echo for transaction formation. Thus, each echo is successively in competition with the decision to form no transaction, not with the other echos. The emitter evaluates echos propagating back from small space-time intervals before proceeding to those echoes from larger intervals. In this way, the time structure of the future universe is in some sense built into the transaction selection process.

The hierarchy assumption resolves the problem that Maudlin had pointed out. The echo from $A$ is considered first, and only if it is rejected is the echo from $B$ considered. The rejection of the $A$ detection triggers the movement of $B$, which then is in place to intercept the left-going wave and produce the echo received and selected by the source. In both cases, the possibility of no detection (and no emission) is a competing possibility. The 50 % probability in the setup is a statement of the relative probabilities of the two detections, and does not include the probability that no particle will be emitted at all in the time interval of the experiment. We note that the physics literature contains several other analyses and resolutions of Maudlin's *gedankenexperiment* [37–40], none of which is identical to the analysis presented here.

Maudlin also claimed that the Transactional Interpretation was deterministic rather than stochastic because it provided no mechanism for randomness. This claim too is incorrect. The intrinsic randomness of the Transactional Interpretation comes in the third stage of transaction formation, in which the emitter, presented with a sequence of retarded/advanced echoes that might form a transaction, hierarchically and randomly selects one (or none) of these as the initial stage of transaction formation, as described in Sect. 5.5. Mead's TI-based mathematical analysis of a quantum jump [41], discussed there, describes a process in which the perturbations between emitter and absorber create a frequency-matched pair of unstable dipole resonators that either exponentially avalanche to a full-blown transaction with the transfer of energy or else disappear due to boundary conditions when a competing transaction forms. In a universe full of particles, this process does not occur in isolation, and both emitter and absorber will be bombarded with perturbations from other systems that can randomly drive the instability in either direction. This is the source of randomness in the Transactional Interpretation.

In this context, it is interesting that Boisvert and Marchildon [40] have suggested that if one assumes determinism and a block universe, the hierarchy described above is not needed. We, however, prefer hierarchy and randomness to determinism, as discussed in Sect. 9.2.

## 6.15 The Afshar Experiment (2003)*

The Afshar experiment [42] shows that, contrary to some of Niels Bohr's pronouncements about complementarity and wave particle duality, it is possible to see the effects of wave-like behavior and interference, even when particle-like behavior is being directly observed. In Bohr's words [43]: " … we are presented with a choice of either tracing the path of the particle, or observing interference effects, … we have to do with a typical example of how the complementary phenomena appear under mutually exclusive experimental arrangements." In the context of a two-slit experiment, Bohr asserted [44] that complementarity in the Copenhagen Interpretation dictates that "the observation of an interference pattern and the acquisition of which-way information are mutually exclusive."

The Afshar experiment, shown in Fig. 6.23 was first performed in 2003 by Shariar S. Afshar and was later repeated while he was a Visiting Scientist at Harvard. It used two pinholes in an opaque sheet illuminated by a laser. The light passing through the pinholes formed an interference pattern, a zebra-stripe set of maxima and zeroes of light intensity that were recorded by a digital camera. The precise locations of the interference minimum positions, the places where the light intensity went to zero, were carefully measured and recorded.

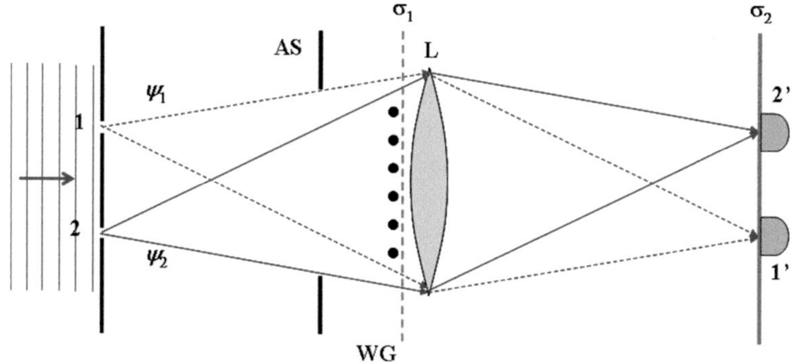


**Fig. 6.23** In the Afshar experiment, a version of Wheeler's delayed-choice experiment (Sect. 6.7) is modified by placing vertical wires (WG) at the locations at which the interference pattern has interference minima on screen $\sigma_1$. High transmission of light through the system when the wires are present and $\sigma_1$ is absent implies that the interference pattern is still present, even when which-way information is available from the downstream detectors $1'$ and $2'$

Behind the plane where the interference pattern formed, Afshar placed a lens that formed an image of each pinhole at a second plane. A light flash observed at image $1'$ on this plane indicated unambiguously that a photon of light had passed through pinhole 1, and a flash at image $2'$ similarly indicated that the photon had passed through pinhole 2. Observation of the photon flashes therefore provided particle-path which-way information, as described by Bohr. According to the Copenhagen Interpretation, in this situation all wave-mode interference effects must be excluded.

However, at this point Afshar introduced a new element to the experiment. He placed one or more vertical wires at the previously measured positions of the interference minima. In such a setup, if the wire plane was uniformly illuminated the wires absorbed about 6 % of the light. Then Afshar measured the difference in the light intensity received at the pinhole image detectors with and without the wires in place.

We are led by the Copenhagen Interpretation to expect that when which-way information is obtained the positions of the interference minima should have no particular significance, and that the wires should intercept 6 % of the light, as they do for uniform illumination. However, what Afshar observed was that the amount of light intercepted by the wires is very small, consistent with 0 % interception. This implies that the interference minima are still locations of zero intensity and that the wave interference pattern is still present, even when which-way measurements are being made. Wires that are placed at the zero-intensity locations of the interference minima intercept no light. This observation would seem to create problems for the complementarity assertions of the Copenhagen Interpretation. Thus, the Afshar experiment is a significant quantum paradox.

The Transactional Interpretation explains Afshar's results as follows: The initial offer waves pass through both slits on their way to possible absorbers. At the wires, the offer waves cancel in first order, so that no transactions to wires can form, and no photons can be intercepted by the wires. Therefore, the absorption by the wires should be very small ($\ll$6 %) and consistent with what is observed. This is also what is predicted by the QM formalism. The implication is that the Afshar experiment has revealed a situation in which the Copenhagen Interpretation has failed to properly map the standard formalism of quantum mechanics.

We note that the many-worlds interpretation of quantum mechanics [29, 30] asserts that interference between its "worlds" (e.g., paths taken by particles) should not occur when the worlds are quantum-distinguishable. Therefore, the Many-Worlds interpretation would also predict that there should be no interference effects in the Afshar experiment. Thus, the Many-Worlds interpretation has also failed to properly map the standard formalism of quantum mechanics.

## 6.16 Momentum-Entangled 2-Slit Interference Experiments (1995–1999)

The Freedman–Clauser and quantum-eraser experiments described in Sects. 6.8 and 6.11 above use conservation of angular momentum and the entanglement of the polarization states of a photon pair to demonstrate EPR correlations and switchable interference patterns. Although the entanglement of linear polarization is a very convenient medium for EPR experiments and Bell-inequality tests, in many ways the alternative offered by momentum-entangled EPR experiments provides a richer venue, and we will consider some of these here.

### *6.16.1 The Ghost-Interference Experiment (1995)**

Perhaps the earliest example of a momentum-entangled EPR experiment is the 1995 "ghost interference" experiment of the Shih Group at University of Maryland Baltimore County [45]. Their experiment is illustrated in Fig. 6.24.

Here a nonlinear BBO ($\beta$-$BaB_2O_4$) crystal pumped by a 351 nm argon-ion laser produces co-linear pairs of momentum-entangled 702 nm photons, one (*e* or extraordinary) polarized vertically and the other (*o* or ordinary) polarized horizontally. These are directed to separate paths by a polarizing beam splitter (PBS).

The experimenters demonstrated that when the pair of photons is examined in coincidence, passing the *e*-photon through a double slit system before detection at

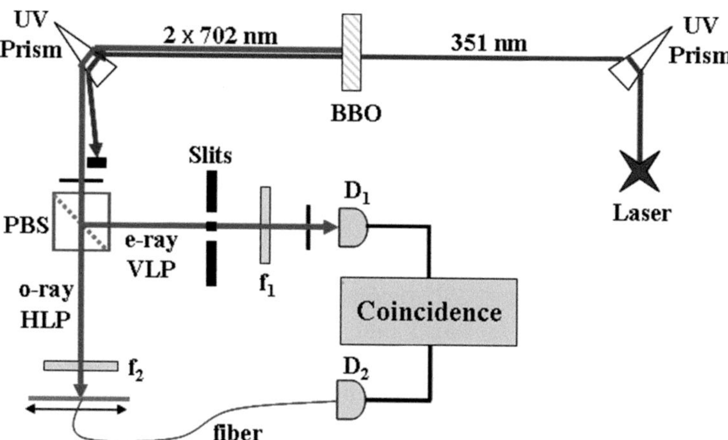


**Fig. 6.24** The "ghost interference" experiment of the Shih Group/UMBC. A BBO crystal produces a momentum-entangled photon pair with orthogonal polarizations, which are split and sent along two paths. Covering one slit in the *e* path switches off the 2-slit interference pattern observed in the *o* path

$D_1$ produced either (1) a "comb" 2-slit interference distribution or (2) a "bump" diffraction distribution in the position $X_2$ of the *e*-photon detected at $D_2$, depending on whether (1) both slits were open so the *o*-photon could take both paths through the slits or (2) one of the slits was blocked, so that which-way information was obtained about the path of the *e*-photon.

Thus, one can make the interference pattern of the *o*-photon observed at $D_2$ appear or disappear, depending on what is done to the *e*-photon. If the *e*-photon is made to exhibit particle-like behavior by passing through only one slit, the *e*-photon also exhibits particle-like behavior. If the *e*-photon is made to exhibit wave-like behavior by passing through both slits and interfering, the *o*-photon also exhibits the wave-like behavior of an interference pattern. This suggests a paradox: that in a system with a momentum-entangled photon pair, a nonlocal signal might be sent from one observer to another by controlling the presence or absence of an interference pattern. To send such a signal, however, one would have to be able to see the interference in singles, without a coincidence with detection of the other member of the photon pair. The Shih Group reported that no interference pattern was observed in singles. See Chap. 7 for a detailed discussion of the possibility of nonlocal signaling.

The Transactional Interpretation explains the ghost interference effect as follows: the *e* and *o* photon are momentum entangled, so that any transaction involving them must conserve transverse momentum, because the transverse momenta of the pair must add up to the near-zero transverse momentum of the pump photon that produced them. Therefore, if the *e* photon deviates to the right of its beam center-line, the *o* photon must deviate a corresponding amount to the left of its beam center-line. Therefore, a two-photon transaction involving detection at $D_1$ of *e* offer waves that have passed through both slits must be accompanied by detection at $D_2$ with *o* offer waves matching the two paths of the *e* waves. Both sets of offer waves will coherently interfere and produce two-slit interference patterns. However, if one of the slits is blocked, the new transaction that forms will involve only one path for each photon, and no two-slit interference will be observed.

### 6.16.2 *The Dopfer Experiment (1999)**

Another momentum-entangled EPR experiment was the 1999 PhD thesis of Dr. Birgit Dopfer at the University of Innsbruck [46], performed under the direction of Prof. Anton Zeilinger. The Dopfer experiment is illustrated in Fig. 6.25.

Here a nonlinear $LiIO_3$ crystal pumped by a 351 nm laser produces momentum-entangled pairs of 702 nm photons and selects pairs that emerge from the crystal at angles of 28.2° to the right and left of the pump axis. The lower photon in the diagram passes through 2-slit system $S_1$ and is detected by single-photon detector $D_1$. The upper photon passes through a lens of focal length $f$ and is detected by single-photon detector $D_2$. The system geometry is arranged so that the distance from $S_1$ to the crystal plus the distance from the upper lens to the crystal add to a total distance of

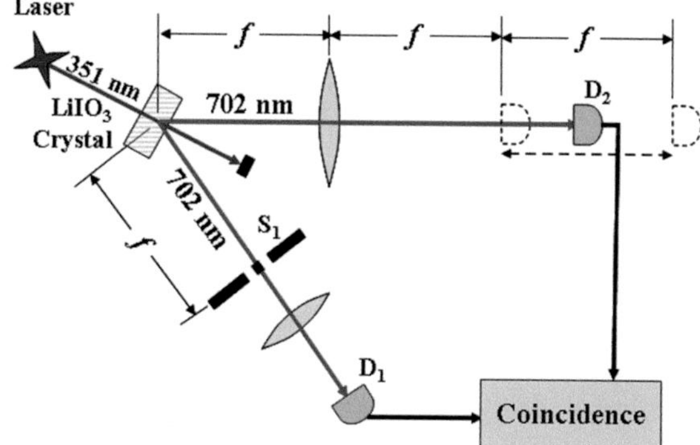


**Fig. 6.25** The Dopfer experiment of the Zeilinger Group/Innsbruck. Here the distance $2f$ (twice the focal length) is the "bent" distance from the lens to the slits $S_1$, and also the distance from the lens to the extreme position of detector $D_2$

$2f$. Beyond the upper lens, detector $D_2$ can be positioned either (Case 1) at a distance of $2f$ from the lens or (Case 2) at a distance of $f$ from the lens. It is observed that for Case 2 an interference pattern is observed at $D_2$, while for Case 1 there is no interference pattern, but only a broad aperture-diffraction distribution.

Dopfer demonstrated that for Case 1, the position distribution measured by detector $D_2$ showed two sharp spikes, which were interpreted as "ghost" images of the slits at $S_1$ that provided which-way information. The slit-lens-detector geometry was such as to produce a 1:1 image, and momentum entanglement caused a right-going photon in the lower system to be mirrored by a left-going photon in the upper system. Thus, in Case 1 detector $D_2$ in effect was measuring which path the lower photon took through the slit system $S_1$ and forcing particle-like behavior in both photons that suppressed the two-slit interference pattern.

In Case 2 the distributions measured by detectors $D_1$ and $D_2$ were both two-slit interference patterns. Detector $D_2$ was placed in the "circle of confusion" region of the lens where no image was formed and virtual rays from both slits would overlap, resulting in interference. Thus, in Case 2 both photons of the entangled pair exhibited wave-like behavior and formed interference patterns.

Therefore, one can make the interference pattern at detector $D_1$ appear or disappear, depending on the location of detector $D_2$. Again, this suggests that in a system with a momentum-entangled photon pair, a nonlocal signal might be sent from one observer to another by controlling the presence or absence of an interference pattern.

The Transactional Interpretaton handles the Dopfer experiment in the same way as the ghost interference experiment discussed in Sect. 6.16.1. When which-way information is available, a 3-vertex transaction can form between $D_1$, the crystal, and only one of the slit images at $D_2$. When $D_2$ is in the circle-of-confusion region, a 3-vertex transaction can form between $D_1$, the crystal, and any point on the interference pattern at $D_2$. Thus, the interference pattern at $D_1$ is "switchable", depending on the distance of $D_2$ from the lens.

The question has been raised [47] of whether this switchable-interference-pattern behavior can be preserved if the coincidence requirements in these two experiments are removed to facilitate nonlocal signaling. This issue is addressed in Chap. 7.

## 6.17 "Boxed Atom" Experiments (1992–2006)

The Stern–Gerlach (SG) effect for measuring atomic spin was first demonstrated in 1922 [48]. It uses the inhomogeneous magnetic field near a wedge-shaped magnetic pole tip to deflect a beam of atoms (in the SG case, spin-$^1/_2$ silver atoms) either upward or downward, depending on the direction in which the atomic spin was pointing with respect to the vertical axis. The beam splitting is, in effect, a measurement of the spin directions of the atoms along a selected axis perpendicular to the beam. The SG effect is reversible, in the sense that a second inhomogeneous magnetic field can recombine the two beams into a single beam of indeterminate spin direction.

### *6.17.1 The Hardy One-Atom Gedankenexperiment*

In 1992 Lucien Hardy [49, 50] proposed the *gedankenexperiment* shown in Fig. 6.26, which is a modified version of the interaction-free measurement scenario of Elitzur and Vaidmann [28] (see Sect. 6.12) in which their blocking object (or bomb) is replaced by a single spin-$^1/_2$ atom, initially prepared in an X-axis $+^1/_2$ spin-projection,

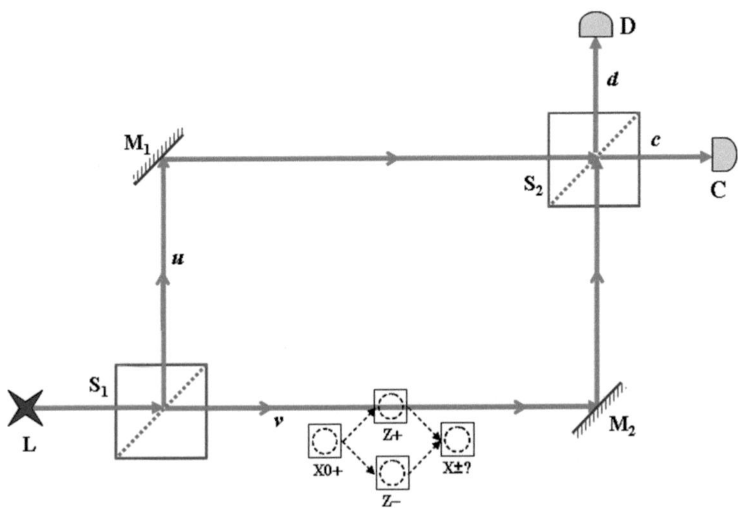


**Fig. 6.26** The Hardy single-atom interaction-free measurement

then Stern–Gerlach separated into one of two spatially separated boxes that momentarily contain the atom in its Z-axis $+{}^1\!/\!_2$ and $-{}^1\!/\!_2$ spin projections, then transmit their contents to be recombined by an inverse Stern–Gerlach process, so that the X-axis projection of the atom can be measured.

The Z-spin $+{}^1\!/\!_2$ box (Z+) is placed directly in one path of a Mach–Zehnder interferometer, so that if the atom is present in that box during photon transit, it has a 100 % probability of absorbing a photon traveling along that arm of the interferometer.[2] After a single photon from light source $L$ traverses the interferometer, the final X-axis spin projection of the atom is measured.

The non-classical outcome of the *gedankenexperiment* is that, for events in which a photon is detected by dark detector $D$, the spin measurement of the atom has a 50 % probability of having an X-axis spin projection of $-{}^1\!/\!_2$, even though the atom had previously been prepared in the $+{}^1\!/\!_2$ X-axis spin state, and the atom had never directly interacted with the photon.

Hardy analyzes the measurement in terms of the Bohm–de Broglie interpretation/revision of quantum mechanics [51] and concludes that the non-classical outcome of the measurement can be attributed to "empty waves", by which he means de Broglie guide waves that have traversed the interferometer along paths not subsequently followed by the single emitted photon. At least four other papers [52–55] have analyzed the Hardy *gedankenexperiment* using alternative QM interpretations that focus on wave function collapse, notably the "collapse" and the "consistent histories" interpretations.

Appendix D.1 provides a detailed analysis of this experiment using the Transactional Interpretation and explains the transfer of non-classical knowledge in terms of the transactional account of the process. In particular, in the case where there is an atom in the $v$ path, it is probed by the offer wave from $L$. When we detect a photon at $D$, (i.e., when a transaction forms between $L$ and $D$), the object has not interacted with a photon (i.e., a transaction has not formed between $L$ and the atom in box $Z+$). However, the atom has been probed by offer waves from $L$, which "feel" its presence and modify the interference balance at the detectors and the spin statistics of the atom. Thus, the Transactional Interpretation gives a simple explanation of the Hardy *gedankenexperiment*.

### 6.17.2 The Elitzur–Dolev Three-Atom Gedankenexperiment

Elitzur and Dolev [56] proposed an elaboration of the Hardy experiment, shown in Fig. 6.27, in which three spin-analyzed Hardy-mode atoms instead of one are placed in boxes intercepting the $v$ interferometer arm. Any of the upper boxes in the $v$ path may be opened, measuring the Z-spin of that atom, or the upper and lower

[2] In the real world, it would be extremely difficult to insure that an isolated trapped atom would intercept a single incident photon with 100 % probability. Thus, Hardy's interesting proposal is doomed to remain a *gedankenexperiment*.

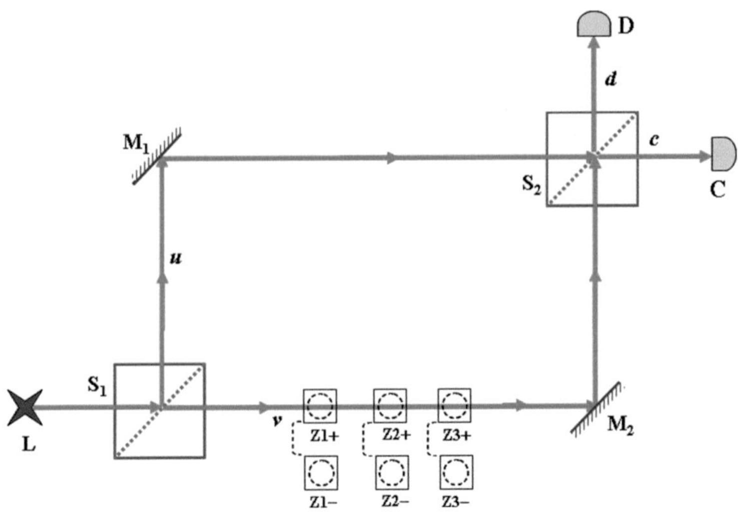


**Fig. 6.27** The Elitzur–Dolev three-atom *Gedankenexperiment*

box contents may be recombined in an inverse Stern–Gerlach procedure and a measurement of the atom's X-spin performed. They do a detailed quantum mechanical analysis of the expected results, which will not be reproduced here.

The most surprising outcome of that analysis is for the case in which a photon is detected in dark detector *D* and one of the three atoms is found to be in the Z-spin $+1/2$ state by opening its upper box on the *v* path and finding the atom to be present there. In this situation, the analysis indicates that the other two atoms *must* be found to have remained in their initially prepared state of X-axis spin $+1/2$. This is true no matter which of the boxes is opened. The other two atoms always remain in their original spin state, unperturbed by the photon, and this result is independent of the order of the atoms along path *v*. Further, a larger number of Hardy-mode atoms could be placed along path *v*, and in that situation it would still be the case that finding one atom with a Z-spin of $+1/2$ would leave all the others unperturbed, even though their wave functions all intercept the photon's possible *v* path before or after the selected atom.

Elitzur and Dolev point out that in this situation, Hardy's empty wave analysis [49, 50] fails, as do all other analyses in the literature [52–55]. They suggest that the Transactional Interpretation might be able to account for the expected non-classical results of the *gedankenexperiment*, and they request that such an analysis, along the lines of the author's previous analysis of interaction free measurements [57], be performed for this *gedankenexperiment*.

The paradoxical aspects of this *gedankenexperiment* arise partly from the fact that there are really four *gedankenexperiments* to be analyzed, one in which all boxes remain closed and three in which one of the boxes is opened. A transactional analysis, as in the case of the AAD *gedankenexperiment* discussed in Sect. 6.10, needs to be performed separately for each of the four experimental configurations.

Following Elitzur and Dolev, we will focus on only those events in which a photon is detected in dark detector $D$, indicating the presence of at least one atom in a $+1/2$ Z-spin state along path $v$. In the box positions, the three atoms may be in $2^3$ or eight different state combinations, which are:

$$| s \rangle \equiv | Z_{1+} \rangle \, | Z_{2+} \rangle \, | Z_{3+} \rangle \tag{6.10}$$

$$| t \rangle \equiv | Z_{1-} \rangle \, | Z_{2+} \rangle \, | Z_{3+} \rangle \tag{6.11}$$

$$| u \rangle \equiv | Z_{1+} \rangle \, | Z_{2-} \rangle \, | Z_{3+} \rangle \tag{6.12}$$

$$| v \rangle \equiv | Z_{1+} \rangle \, | Z_{2+} \rangle \, | Z_{3-} \rangle \tag{6.13}$$

$$| w \rangle \equiv | Z_{1-} \rangle \, | Z_{2-} \rangle \, | Z_{3+} \rangle \tag{6.14}$$

$$| x \rangle \equiv | Z_{1+} \rangle \, | Z_{2-} \rangle \, | Z_{3-} \rangle \tag{6.15}$$

$$| y \rangle \equiv | Z_{1-} \rangle \, | Z_{2+} \rangle \, | Z_{3-} \rangle \tag{6.16}$$

$$| z \rangle \equiv | Z_{1-} \rangle \, | Z_{2-} \rangle \, | Z_{3-} \rangle \tag{6.17}$$

For all of these possibilities except $| z \rangle$, one (or more) of the atoms will block an offer wave on path $v$, suppressing the destructive interference and enabling a possible photon detection at dark detector $D$. In the experimental configuration in which no boxes are opened, the box contents are recombined, and the X-spin states of the atoms are measured, the atom transactions superimposing states $| s \rangle$ to $| y \rangle$. Because of the slight 4 to 3 preference for the Z-axis $+1/2$ spin state in the superposition, the observer will find each atom in the prepared X-spin $+1/2$ with a slightly higher probability than finding other combinations.

Now we assume that box $Z_{2+}$ is opened, and the second atom is found to be in the Z-axis $+1/2$ spin state. This observation is only consistent with offer waves $| s \rangle$, $| t \rangle$, $| v \rangle$, and $| y \rangle$, so the offer wave is a superposition of these states. For the other two atoms, this superposition contains equal amplitudes for Z-axis spin of $+1/2$ and $-1/2$ with no alteration of phase, so if the contents of boxes 1 and 3 are recombined and the X-axis spins measured, the resulting transactions will require both atoms have X-axis spins of $+1/2$, as was originally prepared. If we assume that box $Z_{1+}$ or $Z_{3+}$ is opened instead, we will obtain the same result for the other two atoms. Elitzur and Dolev describe this result as: "In other words, only one atom is affected by the photon in the way pointed out by Hardy, but that atom does not have to be the first one, nor the last; it can be any one out of any number of atoms. The other atoms, whose wave-functions intersect the Mach–Zehnder interferometer arm before or after that particular atom, remain unaffected."

In the context of the Transactional Interpretation, is it really true that only one of the Hardy atoms interacts with the photon? No. The offer waves for the photon detected at $D$ have interacted with all of the atoms, but have done so in such a way as to force all remaining atoms into their original state when one selected atom is found to be in the path-blocking Z-axis $+1/2$ spin state. The allowed composite multi-vertex transactions between transaction vertexes $L$, $D$, $Z_{2+}$, $X_{1+}$, and $X_{3+}$, or the equivalent for other measurement scenarios, produce this result. The Transactional Interpretation has no difficulty in explaining this counter-intuitive result.

### 6.17.3 The Elitzur–Dolev Two-Atom Gedankenexperiment

In 2006 Elitzur and Dolev [56] also proposed another revision of the Hardy experiment, shown in Fig. 6.28, in which spin-analyzed atoms are placed in boxes intercepting both of the Mach–Zehnder interferometer arms. In the interest of brevity, we will leave it as an exercise for the reader to repeat the detailed analysis of the previous sections. Instead, we observe that this is a direct extension of Hardy's *gedankenexperiment*, except that there are now nine transaction vertices: $L$, $C$, $D$, $X10_+$, $Z1_+$, and $X1_\pm$, $X20_+$, $Z2_-$, and $X2_\pm$. The same boundary conditions discussed above apply to these photon and atom vertices.

Again focusing on non-classical events involving photon detection at dark detector $D$, we see that such detection requires that an atom in box $Z1_+$ blocks the $v$ path or an atom in box $Z2_-$ blocks the $u$ path, but not both. Because of the superposition of the two possibilities, the atom detectors $X1_\pm$ and $X2_\pm$ will measure a probability $\frac{3}{4}$ of observing the previously prepared spin state and a probability $\frac{1}{4}$ of observing the opposite spin state.

Elitzur and Dolev, in considering the possibility of a transactional analysis of this *gedankenexperiment*, stated "Once the interaction time with the atom is over and no absorption occurred, two facts need to be addressed: (1) The wave-function is radically changed, now giving probability 1 that the photon is in the other

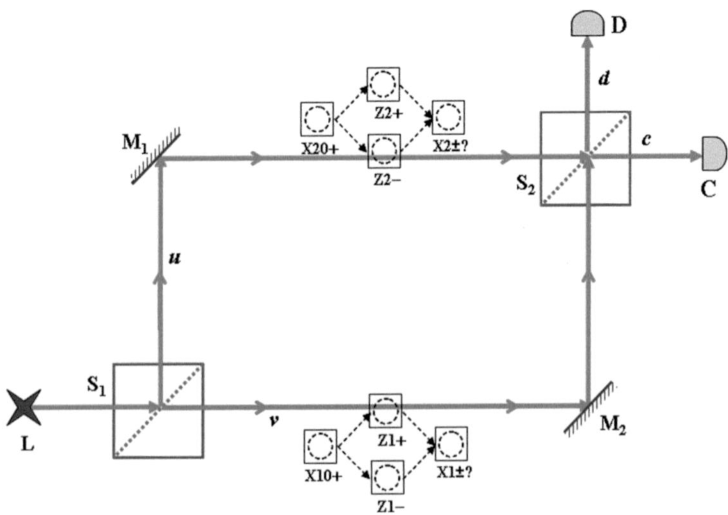


**Fig. 6.28** The Elitzur–Dolev two-atom *Gedankenexperiment*

Mach–Zehnder interferometer path and (2) One can remove the second beam splitter and the detectors in a 'delayed choice experiment' fashion, preventing the final interaction altogether. Now, if the TI insists that the 'confirmation waves' from the atom and from the detectors arrive to the source together, the resulting account cannot properly handle these intermediate stages". They suggested that some recursive "go-back-and-start-again" structure to the transactions would be required in this case, in the form of a "cancellation wave". That, however, is not a part of the Transactional Interpretation.

The Elitzur and Dolev account contains a misperception of how the Transactional Interpretation should be applied to multi-vertex quantum events. The emitters receive the ensemble of confirmation waves from potential absorbers and make a hierarchical probabilistic choice, based on their amplitudes, with possible transactions from "near" (small space-time interval) absorbers (in this case, the atoms) confirmed or rejected before transactions involving "far" absorbers (in this case, the detectors) are considered (see Sect. 5.6.). Because of the hierarchy in transaction formation, there is no need for any recursive procedure or the invocation of cancellation waves. Further, probabilities are calculated only for completed transactions. Mid-process reevaluation of probabilities is a peculiarity of the knowledge interpretation (see Sect. 6.6) and is not a part of the TI. Any change in the configuration, such as removal of a beam splitter, represents a separate and distinct configuration from the one being analyzed and must be analyzed as a separate system with its own set of possible transactions that are also selected hierarchially. Thus, the TI, when properly applied, meets the Elitzur–Dolev challenge with no problems.

### 6.17.4 The Time-Reversed EPR Gedankenexperiment

Elitzur and Dolev went on to propose a variation of the two-atom experiment [56, 58] (Sect. 6.17.3) in which the initial light source and splitter are replaced by two distant light sources $S_1$ and $S_2$ that are synchronized to coherently produce light of identical wavelengths. This is shown in Fig. 6.29. The sources are of very low intensity, so that, on the average, only one photon is emitted during a given time interval, and it may be emitted from either source. The probability of emission for each source is $P_\gamma \ll 1$. We note that arranging for two widely separated coherent sources of light can be implemented by sending the sources ultra-fast synchronizing pulses, as discussed in Sect. 6.19 on entanglement swapping.

The two beams cross at a 50:50 beam splitter *BS* and are then detected by single-photon detectors *C* and *D*. The path lengths are arranged so that the coherent waves from the two sources destructively interfere at *D* and constructively interfere at *C*. Operationally, this situation differs from the two-atom experiment (Sect. 6.17.3) only in that there is a possibility that two simultaneous photons may be detected or absorbed by the atoms. The probability of such a coincidence is $P_\gamma^2 \ll 1$, which is so small that it can be completely neglected.

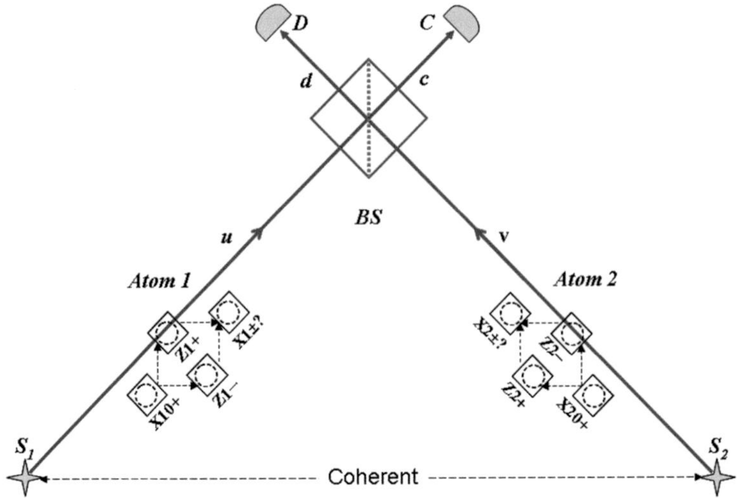


**Fig. 6.29** The time-reversed EPR *Gedankenexperiment*. Weak coherent sources $S_1$ and $S_2$ send light through Hardy atoms to a beam splitter (*BS*) and detectors *C* (constructive) and *D* (destructive). For detections at *D*, EPR Bell-state correlations are observed

As before, the atoms initially have a positive spin along the $x$-axis, are SG separated based in their spin projection on the $z$-axis, and sent to two intermediate boxes, with a probability of 50 % that the atom may reside in each box at the time a photon passes through it. If a box is occupied by an atom when a photon passes through, there is a 100 % probability that the photon will be absorbed, leaving the atom in an excited state, with no light waves transmitted any further along the path.

Elitzur and Dolev focus exclusively on events in which detector *D* detects a photon (and *C* does not). They argue that, in the sense of interaction-free measurements (Sect. 6.12), one of the two boxes, but not both, must have been occupied by an atom just before the detection. This means that the wave function for detection at *D* was:

$$\mid D\rangle = {}^1\!/\!_4\,(\mid Z_{1+}\rangle \mid Z_{2+}\rangle + \mid Z_{1-}\rangle \mid Z_{2-}\rangle), \tag{6.18}$$

where $\mid Z_{(1,2)\pm}\rangle$ indicates the Z-axis spin projection of atom 1 or 2. Thus, the two atoms, which have never interacted, are entangled in a full-blown EPR Bell state in which their Z-axis spin projections must match. (see Eq. 2.1):

In other words, for *D* detections subsequent tests of Bell's inequality performed on the two boxed atoms, e.g., by rotating the axis of one SG $x$-axis recombiner with respect to the other, will show the same Bell inequality violations observed in EPR tests like the Freedman–Clauser experiment (Sect. 6.8), and indicating that the spin value of each atom depends on the choice of spin direction measured for the other atom, no matter how distant.

Unlike the more conventional EPR experiments, in which the particles are entangled in a Bell state because they have interacted earlier, here the only common event between the two atoms lies in their *future*. One might argue that the atoms are measured only after the photon's interference and detection, hence the entangling event still resides in the measurements' past. However, all three events, namely, the photon's interference and the two atoms' measurements, can be performed at large spacelike separations. In that case, by suitable choice of reference frame, the entangling event may be made to reside either in the measurements' past or its future. Thus, this *gedankenexperiment* is truly a time-reversed EPR experiment.

The Transactional Interpretation analysis of this experiment is similar to previous analyses. One might think that a weak source might only occasionally send out offer waves. However, the proper TI view is that a weak source should continuously emit very weak offer waves, which only occasionally result in the formation of a transaction. Because of destructive interference that has been arranged, detector *D* can receive these offer waves only for the situation in which one of the paths is blocked by the presence of an atom in a Hardy box. For such offers, it returns confirmation waves to the appropriate source, a transaction can form, and a photon can be transferred from that source to detector *D*. Such transactions only occur when one Hardy box is empty and the other occupied, leading to the selection of only Bell-state offer wave functions as those capable of forming a transaction to detector *D*. Thus, the TI easily accounts for the curious time-reversed EPR results.

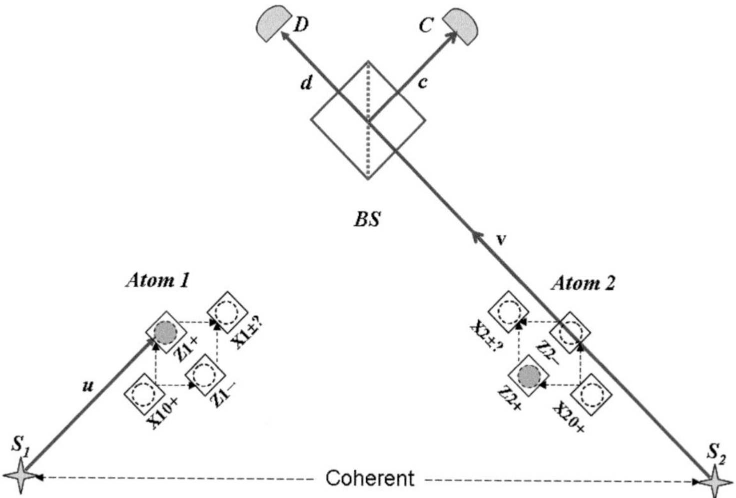


**Fig. 6.30** The Elitzur–Dolev Quantum Liar Paradox *Gedankenexperiment*

### *6.17.5 The Quantum Liar Paradox*

Finally, Elitzur and Dolev (ED) consider the implications of the time-reversed EPR experiment for the logic of the spin orientation of one atom affecting the spin orientation of another, when they have neither interacted with each other nor with a passing photon.

In the situation illustrated in Fig. 6.30 we find that after detection at $D$, discovering, for example a spin-up atom in box 1, has the following implications:

1. Atom 1 is positioned in the intersecting box $Z1^+$.
2. It has not absorbed any photon.
3. Still, the fact that the spin of Atom 2 is affected by the position of Atom 1 means that *something* has traveled the path blocked by Atom 1. To prove that, let an opaque object be placed on path $u$ after Atom 1. No EPR nonlocal correlations will be observed.

ED argue that the very fact that one atom is positioned in a location that seems to preclude its interaction with the other atom is affected by that other atom. They say that this is logically equivalent to the statement "this sentence has never been written." They state that they are unaware of any other quantum mechanical experiment that demonstrates such an inconsistency.

The Transactional Interpretation explains this paradox by observing that Atom 1 is probed by an offer wave from $S_1$ that it blocks, even if no transaction occurs between $S_1$ and Atom 1. The absence of this offer wave beyond splitter $BS$, because it stops at Atom 1, prevents destructive interference in path $d$ and allows a transaction between $S_2$ and $D$ to form, provided Atom 2 does not block that path. The transactional handshake between $S_2$ and $D$ is only possible because path $u$ is blocked and path $v$ is open. Placing an opaque object on path $u$ would allow transactions between $S_2$ and $D$ with Atom 1 in both possible spin orientations and would destroy the Bell-state offer wave selection and EPR correlations. This, however, does *not* prove that "something has traveled the path blocked by Atom 1" as claimed by ED. Thus, the TI has no problem in explaining the Quantum Liar paradox and its underlying logic. We note that Kastner [37] and Boisvert and Marchidon [40] have also published somewhat different Transactional Interpretation analyses of the quantum liar paradox.

## 6.18 The Leggett–Garg Inequality and "Quantum Realism" (2007)*

Noble Laureate Anthony J. Leggett of the University of Illinois has demonstrated that by focusing on the falloff of correlations with *elliptical* polarization (mixtures of circular + linear polarization), rather than on the linear polarization of the Bell Inequality EPR experiments, one can compare the predictions of quantum mechanics with a class of nonlocal realistic theories [59–61]. The resulting Leggett–Garg

Inequalities can be used in the same way as the Bell Inequalities, but to test nonlocal realism instead of local realism.

A group of experimentalists at Anton Zeilinger's Institute for Quantum Optics and Quantum Information (IQOQI) in Vienna have performed an EPR experiment that is a definitive test of the Leggett–Garg Inequalities [62]. They show that in EPR measurements with elliptically polarized entangled photons, the Leggett–Garg Inequalities in two observables are violated by 3.6 and by 9 standard deviations. This is interpreted as a statistically significant falsification of the whole class of nonlocal realistic theories studied by Leggett.

The group summarizes the implications of their results with this statement: "We believe that our results lend strong support to the view that any future extension of quantum theory that is in agreement with experiments must abandon certain features of realistic descriptions." In other words, quantum mechanics and reality appear to be incompatible and have parted company.

Is the case against objective reality truly so strong? To answer this question, we must examine in more detail the nonlocal realistic theories that Leggett studied. This class of theories assumes that when entangled photons emerge from their emission source, they are in a *definite but random state of polarization*. That is Leggett's definition of "realism". It is well known from the work of Furry [63, 64] that when that assumption (and no other) is made, one does not observe the quantum mechanical prediction of Malus's Law for the correlations of the photon pair.

However, Leggett cures that problem by assuming an unspecified nonlocal connection mechanism between the detection systems that fixes the discrepancy. In effect, the two measurements talk to each other nonlocally in such a way that the detected linearly polarized photons obey Malus' Law and produce the same EPR polarization correlations predicted by quantum mechanics. Leggett then shows that this nonlocal "fix" cannot be extended into the realm of elliptical polarization and that quantum mechanics and this type of nonlocal realistic theories give differing predictions for the elliptical polarization correlations. In other words, the "reality" that is being tested is whether the photon source is initially emitting the entangled photons in a *definite but random state of polarization*. It is this version of reality that has been falsified by the IQOQI measurements.

We can clarify what is going on in these experimental tests by applying the Transactional Interpretation to these Leggett–Garg Inequality tests. From the point of view of the TI and standard quantum mechanics, Leggett's assumption that the entangled photons are emitted in definite states of polarization is simply wrong. The "offer wave" for each photon that emerges from the source includes all possible polarization states. These offer waves travel to downstream detectors, and time-reversed "confirmation waves" travel back up the time-stream to the source, arriving at the instant of emission. As was illustrated for the Freedman–Clauser experiment in Fig. 6.9, a three-way transaction then forms between the source and the two detections that matches the confirmation waves to a mutually consistent overall state that satisfies appropriate conservation laws (in this case, conservation of angular momentum). The final result is a completed transaction with the two photons in definite states, but this definite state was not present in the initial emission of the offer waves, and that is the part of the process described in detail by the wave-mechanics formalism of quantum mechanics. We note that the TI does not in itself make any predictions about the linear or elliptical polarization correlations of the entangled photon pair. It only describes the quantum formalism that is making the predictions that the IQOQI group has observed to be consistent with their experiment, but it clarifies what is going on in those predictions.

Does this mean that the TI (and the quantum formalism it describes) are not "realistic", i.e., inconsistent with an objective reality that is independent of the observer's choice of measurements? I don't think so. The transactions that form in quantum processes arise from a "handshake" between the past and future across space-time, but they are not specifically the result of measurements or observer choices. The latter are only a small subset of the transactions that form as the universe evolves in space-time. The message of the Leggett–Garg Inequality tests, from the point of view of the TI, is that the assumption of emission in a definite polarization state is too restrictive. I would argue that initial emission without a definite polarization state is perfectly consistent with objective reality and is consistent with the quantum formalism. It is just that reality is not fixed by the initial offer wave and does not become "frozen" until the transaction is formed.

The TI description of the quantum formalism is both realistic and nonlocal, in at least some definitions of those terms, and it is completely consistent with the IQOQI results. To put it another way, Leggett has set up a straw man that has been demolished by the IQOQI tests, but that is only an indication that his version of "realism" is too naïve. And this theory and experiment can be viewed as another demonstration of the value and power of the TI in understanding the peculiar predictions and intrinsic weirdness of quantum mechanics.

## 6.19 Entanglement Swapping (1993–2009)*

In conventional telecommunication systems, transmission of signals over significant distances uses "repeaters", devices that receive incoming signals, reshape and amplify them, and send them along to the next repeater station. If quantum information contained in photon entanglement needs to be sent over long distances and is to be "repeated" in the same way, there is a significant problem in how such a "quantum repeater" might operate while preserving entanglement.

The solution to this problem seems to be *entanglement swapping*, first proposed in 1993 [65]. Briefly, this is accomplished by mixing an entangled photon taken from each of two synchronized entangled photon sources and performing Bell-state measurements on the mixed pair. The consequence is that the two unmixed outgoing photons from the two sources will be entangled, and the type of entanglement can be selected by coincidences with the detected photons of the Bell-state measurements. Interestingly, the technique produces photons that are entangled, even though they have never interacted and they originate in separate locations. Somehow, because

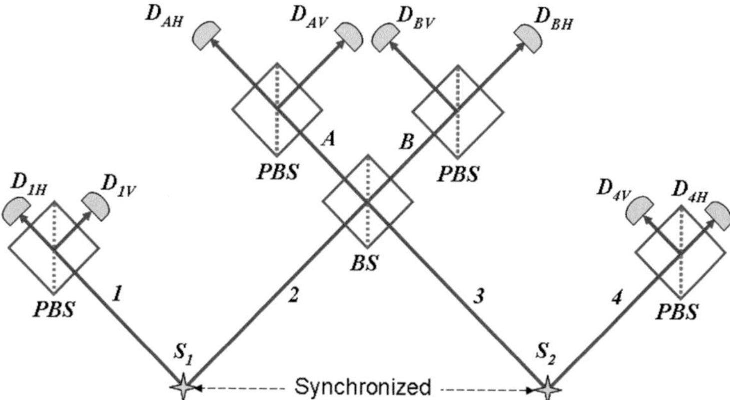


**Fig. 6.31** An experimental configuration for producing entanglement swapping. Synchronized sources $S_1$ and $S_2$ produce polarization-entangled pairs (*1, 2*) and (*3, 4*). Photons *2* and *3* are mixed at *BS* and their Bell-state detected by polarimeters *A* and *B*. Photons *1* and *4*, which have never interacted, are entangled

their entangled twins interact, the entanglement is "swapped" to the non-interacting pair.

Figure 6.31 shows an experimental configuration [66] for producing this entanglement swapping. Synchronized entangled two-photon sources $S_1$ and $S_2$ produce polarization-entangled pairs (1, 2) and (3, 4). The polarization-entanglement is such that if one photon is vertically polarized ($V$) the other is horizontally polarized ($H$), and *vice versa*. Photons 2 and 3 are mixed and their Bell-state detected by polarimeters $A$ and $B$ with detectors $D_{AH}$, $D_{AV}$, $D_{BH}$, and $D_{BV}$, using Hong–Ou–Mandel interference [67]. Only two photons are detected, but they may be detected by any pair of polarimeter detectors. Therefore, in principle the two detections may be in any of 6 combinations: $(H_A, V_A)$, $(H_A, H_B)$, $(H_A, V_B)$, $(V_A, H_B)$, $(V_A, V_B)$, or $(H_B, V_B)$. The result of this is that photons 1 and 4 are entangled so that, in coincidence with the polarimeter detections, they are projected into a Bell state (see Eqs. 2.1 and 2.2) that depends on the detection combination. They will be in the Bell state $\frac{1}{\sqrt{2}}(|H_1V_4\rangle - |V_1H_4\rangle)$ if they are in coincidence with $(H_A, V_B)$ or $(V_A, H_B)$, and they will be in the Bell state $\frac{1}{\sqrt{2}}(|H_1V_4\rangle + |V_1H_4\rangle)$ if they are in coincidence with $(H_A, V_A)$ or $(H_B, V_B)$. The polarimeter combinations $(H_A, H_B)$ and $(V_A, V_B)$ do not occur because the sources are set for opposite-polarization entanglement.

The entanglement of 1 and 4 can be eliminated by removing the beam splitter *BS*, so that there is no mixing and the *A* polarimeter measures the polarization of 3 and the *B* polarimeter measures the polarization of 2. In this case 1 and 4 are in unentangled product states. Curiously, the path to the *A* and *B* polarimeters can be made much longer than the 1 and 4 paths, so that the decision of whether 1 and 4 should be entangled or not can be made *after* these photons have already been detected.

So far, entangled photon transmission has been limited to distances on the order of 100 km [68–70]. To increase transmission distances beyond this level, quantum repeaters are needed. The entanglement-swapping configuration discussed here has been tested [66] and found to give high quality entanglement that would be suitable for use in a quantum repeater. Imagine that $S_1$ is the beginning of a quantum transmission line, and path 1 is short while 2 is very long, and that ultra-fast synchronizing pulses and the output of the 1 polarimeter are sent on a path parallel to 2. Source $S_2$ is synchronized with $S_1$ and produces photons 3 and 4. Photon 3 is mixed with 2 and analyzed. Photon 4 becomes the "repeated" version of 2 and is sent along the line, along with the polarimeter outputs and the synchronizing pulse. This process can be repeated at suitable transmission length intervals indefinitely, leading, at least in principle, to the development of large scale quantum communication networks.

The Transactional Interpretation makes it easy to understand what is going on in entanglement swapping. One must consider the formation of a transaction for each configuration in which one $H$ and one $V$ photon are detected at polarimeters 1 and 4. Without going into the details of the transactions involved, it is easy to see that, for example, a single transaction that involves dual emissions at $S_1$ and $S_2$ and detections at $D_{1H}$ and $D_{4V}$ will require either matching detections at $D_{AH}$ and $D_{BV}$ or detections at $D_{AV}$ and $D_{BH}$. These involve a network of offer and confirmation waves linking the vertices of the transaction. Similarly, it is easy to see that transactions involving detections at $D_{1H}$ and $D_{4H}$ cannot be completed if the $A$ and $B$ polarimeters detect one $H$ and one $V$ photon, so detections inconsistent with the (1, 4) entanglement are forbidden. Thus, by selecting HV coincidences in the A and B detectors, the wave function for photons 1 and 4 is a Bell state, and the entanglement has been transmitted. One can also see that if the beam splitter *BS* is removed, only separate and independent transactions will form between polarimeters 1 and $B$ and between polarimeters $A$ and 4, so there will be no (1, 4) entanglement.

## 6.20 Gisin: Neither Sub- nor Superluminal "Influences"? (2012)

The Gisin group [71] has examined the nonlocality of quantum mechanics from another direction. They consider Bell-type EPR experiments in which entangled pairs of photons are given entangled polarizations by the emission process (through angular momentum conservation) and their polarization states are measured in some selected polarization basis (H/V linear, $\pm 45°$ linear, or L/R circular) by downstream detectors. Quantum mechanics requires that whenever the detection bases of two such measurements match, the measured values must also match.

The authors assume that they can replace orthodox quantum mechanics by some unspecified semi-classical process in which the "causal influences" have a well defined propagation velocity and travel between measurements to insure that the polarization correlations match. It has already been well established through the work of J. S. Bell and others that any such causal influences traveling at velocities less than or equal to the speed of light cannot account for the EPR correlations observed in Bell-type EPR experiments. The authors extend consideration to include causal influences traveling at velocities *greater than the speed of light*. They show that causal influences traveling at velocities greater than the speed of light can indeed account for EPR correlations, but the assumption of superluminal influences carries with it the inevitable consequence that signaling between observers at the superluminal speed of the causal influences becomes possible.

Special relativity (see Sect. 7.2) forbids such signaling at any well-defined superluminal speed because its existence would allow the discovery of a preferred reference frame and would destroy the even-handedness with which relativity treats all inertial reference frames. Thus, the authors concluded that no semiclassical explanation of quantum nonlocality and EPR correlations is possible, even when superluminal causal influences are allowed.

We note that extensions of the many-worlds interpretation have attempted to deal with quantum nonlocality by hypothesizing a traveling "split" between worlds, i.e., universes, that originates at the site of one measurement and propagates to the sites of other measurements, in order to arrange consistent EPR correlations between measurement results. This moving split is just the kind of moving causal influence with a well defined propagation velocity that has been ruled out by the Gisin group's paper.

The work presents a hypothesis that some have seriously entertained and then demonstrates its unacceptable implications. However, the basic approach, one that has been taken by many other works in the physics literature, seems intended to mystify and obscure quantum mechanics and nonlocality rather than to clarify and understand them.

The Transactional Interpretation, which is not referenced or considered in the Gisin group's paper, describes "causal influences", i.e., the wave functions $\psi$ and $\psi^*$ of the emitted entangled photons, as propagating in both time directions along the allowed trajectories of the particles and handshaking to observe conservation laws by building in the observed EPR correlations. The causal influences are not superluminal, but rather retro-causal. Does this causal link imply that superluminal signaling is possible? Not in the sense considered in the Gisin group's paper. The lines of communication for the entangled EPR photons, as described by the Transactional Interpretation, are all along light-like world lines that transform properly under the Lorentz transformations of relativity, favoring no preferred inertial reference frame and remaining completely consistent with special relativity (see Sect. 7.2).

## 6.21 The Black Hole Information Paradox (1975–2015)

Stephen Hawking's 1975 calculations [72] predicting black hole evaporation by Hawking radiation described a process that apparently does not preserve information.

This created the Black Hole Information Paradox, which has been an outstanding problem at the boundary between general relativity and quantum mechanics ever since. Lately, gravitational theorists have focused on pairs of quantum-entangled particles, in part because the particle pair involved in Hawking radiation should be entangled. They have considered ways in which the quantum entanglement might be broken or preserved when one photon of the entangled photon pair crosses the event horizon and enters a black hole.

One recent suggestion is that the quantum entanglement breaks (whatever that means) when the infalling member of the entangled particle pair crosses the event horizon, with each breaking link creating a little burst of gravitational energy that cumulatively create a firewall just inside the event horizon. This firewall then destroys any infalling object in transit [73]. The firewall hypothesis, however, remains very controversial, and there is no apparent way of testing it.

More recently Maldacena and Susskind [74] have suggested an alternative. When two entangled black holes separate, they hypothesize that a wormhole connection forms between them to implement their entanglement. It has even been suggested that such quantum wormholes may link *all* entangled particle pairs. There are, however, problems with this interesting scenario, not the least of which is that such wormholes should have significant mass that is not observed.

The Transactional Interpretation offers a milder, if less dramatic solution to this problem, providing an interesting insight into the Black Hole Information Paradox. One normally thinks that absolutely nothing can break out of the event horizon of a black hole from the inside and escape. However, there is one exception: advanced waves can emerge from a black hole interior, because they are just the time-reverse of a particle-wave falling in. An advanced wave "sees" the black hole in the reverse time direction, in which it looks like a white hole that emits particles. The strong gravitational force facilitates rather than preventing the escape of an advanced wave. Thus, an entangled particle pair, linked by an advanced-retarded wave handshake, have no problem in maintaining the entanglement, participating in transactions, and preserving conservation laws, even when one member of the pair has fallen into a black hole. There is no need for entanglement-breaking firewalls or entanglement-preserving wormholes, just a transactional handshake. Thus, it would seem that the Transactional Interpretation goes some considerable distance toward solving the Black Hole Information Paradox and resolving an issue that divides quantum mechanics and gravitation and providing a mechanism for preserving information across event horizons.

## 6.22 Paradox Overview

In summary, there is a large and growing array of interpretational paradoxes and puzzles arising from the formalism of quantum mechanics and its peculiar properties and behavior. New quantum optics experiments are published every day that demonstrate the intrinsic counter-intuitive weirdness of the quantum world. Heisenberg's knowledge interpretation, a central part of the Copenhagen Interpretation, had been able to deal with some of these problems, but its focus on observer knowledge appears inadequate to deal with systems involving multiple measurements, multiple observers, and multiple choices that may be made in any time sequence.

The emphasis of the knowledge interpretation on the observer and his knowledge has led us into some philosophically deep waters. It is asserting that somehow, the solutions of a simple second-order differential equation relating mass, energy, and momentum have entered the head of an intelligent observer and are describing his state of knowledge about the outside world. It leads to the conclusion that, in some sense, the observer is "creating" the external reality by his choice of observations, choosing to make one member of a pair of conjugate variables "real" at the expense of the other by deciding to measure it. Quantum mysticism based on such observer-created reality has become a popular theme in books that attempt to sensationalize physics for the general reader, finding tenuous and deceptive connections between the Copenhagen brand of quantum mechanics and the dogmas of exotic religions.

Further, the positivism of the Copenhagen Interpretation frustrates our desire to "view" quantum processes and to understand what goes on "behind the scenes" that can lead to such curious and paradoxical behaviors in the quantum world. The Transactional Interpretation provides a straightforward way of resolving these paradoxes and problems and eliminating the need for appeal to observer knowledge. It also provides the tools for visualizing the underlying mechanisms in quantum processes.

## References


1. R.P. Feynman, R.B. Leighton, M. Sands, *The Feynman Lectures*, vol. 3 (Addison-Wesley, Reading, 1965). ISBN: 0201021188
2. Sir G.I. Taylor, Interference fringes with feeble light. Proc. Camb. Philos. Soc. **15**, 114 (1909)
3. G.N. Lewis, The nature of light. Proc. Natl. Acad. Sci. **12**, 22–29 (1926)
4. T.L. Dimitrova, A. Weis, The wave-particle duality of light: a demonstration experiment. Am. J. Phys. **76**, 137–142 (2008)
5. A. Einstein, in *Electrons et Photons - Rapports et Discussions du Cinqui'ème Conseil de Physique tenu, Bruxelles du 24 au 29 Octobre 1927 sous les Auspices de l'Institut International de Physique Solvay* (Gauthier-Villars, Paris 1928)
6. M. Jammer, *The Conceptual Development of Quantum Mechanics* (McGraw-Hill, New York, 1966)
7. E. Schrödinger, Proc. Camb. Philos. Soc. **31**, 555–563 (1935)
8. M. Renninger, Zeitschrift für Physik **136**, 251 (1953)
9. L. de Broglie, *The Current Interpretation of Wave Mechanics* (Elsevier, Amsterdam, 1964)
10. J. von Neumann, *Mathematische Grundlagen der Quantenmechanik* (Springer, Berlin, 1932)
11. E.P. Wigner, in *The Scientist Speculates*, ed. by I.J. Good (Heinemann, London, 1962)
12. W. Heisenberg, *Physics and Beyond* (Harper and Rowe, New York, 1960), pp. 60–62
13. A. Neumaier, Collapse challenge for interpretations of quantum mechanics (unpublished), arXiv:0505172v1 [quant-ph]
14. J.A. Wheeler, in *The Mathematical Foundations of Quantum Mechanics*, ed. by A.R. Marlow (Academic Press, New York, 1978)

15. V. Jacques, E. Wu, F. Grosshans, F. Treussart, P. Grangier, A. Aspect (LCFIO), J.-F. Roch, Experimental realization of Wheeler's delayed-choice *gedankenexperiment*. Science **315**, 966 (2007), arXiv:0610241 [quant-ph]
16. S.J. Freedman, J.F. Clauser, Phys. Rev. Lett. **28**, 938 (1972)
17. J.S. Bell, Physics **1**, 195 (1964)
18. R. Hanbury Brown, R.Q. Twiss, A test of a new type of stellar interferometer on Sirius. Nature **178**(4541), 1046–1048 (1956)
19. J.R. Klauder, E.C.G. Sudarshan, *Fundamentals of Quantum Optics* (Benjamin, New York, 1968)
20. M. Gyulassy, S.K. Kauffmann, L.W. Wilson, Phys. Rev. C **20**, 2267 (1979)
21. J.G. Cramer, G.A. Miller, J.M.S. Wu, J.-H. Yoon, Phys. Rev. Lett. **94**, 102302 (2005), arXiv:0411031 [nucl-th]
22. G.A. Miller, J.G. Cramer, Polishing the lens: I pionic final state interactions and HBT correlations - distorted wave emission function (DWEF) formalism and examples. J. Phys. **G34**, 703–740 (2007)
23. M. Luzum, J.G. Cramer, G.A. Miller, Understanding the optical potential in HBT interferometry. Phys. Rev. C **78**, 054905 (2008)
24. R.L. Pflegor, L. Mandel, Phys. Rev. **159**, 1084 (1967)
25. D.Z. Albert, Y. Aharonov, S. D'Amato. Phys. Rev. Lett. **54**, 5 (1985)
26. Y. Aharonov, P.G. Bergmann, J.L. Lebowitz, Phys. Rev. **134**, B1410 (1964)
27. T.J. Herzog et al., Phys. Rev. Lett. **75**, 3034–3037 (1995)
28. A.C. Elitzur, L. Vaidman, Found. Phys. **23**, 987–997 (1993)
29. H. Everett III, Rev. Mod. Phys. **29**, 454 (1957), see also [30]
30. J.A. Wheeler, Rev. Mod. Phys. **29**, 463 (1957), see also [29]
31. L. Mandel, E. Wolf, *Optical Coherence and Quantum Optics* (Cambridge University Press, Cambridge, 1995)
32. P.G. Kwiat et al., Phys. Rev. Lett. **83**, 4725–4728 (1999)
33. T. Maudlin, *Quantum Nonlocality and Relativity* (Blackwell, Oxford, 1996, 1st edn.; 2002, 2nd edn.)
34. R.E. Kastner, *Understanding Our Unseen Reality: Solving Quantum Riddles* (Imperial College Press, London, 2015)
35. P.J. Lewis, Retrocausal quantum mechanics: Maudlin's challenge revisited. Stud. Hist. Philos. Mod. Phys. **44**, 442–449 (2013)
36. J.G. Cramer, The transactional interpretation of quantum mechanics. Rev. Mod. Phys. **58**, 647–687 (1986)
37. R.E. Kastner, Cramer's transactional interpretation and causal loop problems. Synthese **150**, 1–40 (2006)
38. L. Marchidon, Causal loops and collapse in the transactional interpretation of quantum mechanic. Phys. Essays **38**, 807–814 (2006)
39. R.E. Kastner, On delayed choice and contingent absorber experiments. ISRN Math. Phys. **2012**, 617291 (2012)
40. J-S Boisvert and L. Marchidon, Absorbers in the transactional interpretation of quantum mechanics, arXiv:1207.5230v2 [quant-ph]
41. C. Mead, *Collective Electrodynamics* (The MIT Press, Cambridge, 2000). ISBN: 0-262-13378-4
42. S.S. Afshar, Violation of the principle of complementarity, and its implications. Proc. SPIE **5866**, 229-244 (2005), arXiv:0701027 [quant-ph]
43. N. Bohr, Discussions with Einstein on epistemological problems in atomic physics, in *Albert Einstein: Philosopher-Scientist*, ed. by P. Schilpp (Open Court, Peru, 1949)
44. N. Bohr, *Atti del Congresso Internazionale dei Fisici Como, 11-20 Settembre 1927*, vol. 2 (Zanchelli, Bologna, 1928), pp. 565–588
45. D.V. Strekalov, A.V. Sergienko, D.N. Klyshko, Y.H. Shih, Phys. Rev. Lett. **74**, 3600–3603 (1995)
46. B. Dopfer, Ph.D. thesis, University, Innsbruck (1998, unpublished)

47. R. Jensen, *Proceedings of STAIF 2006*. AIP Conf. Proc. **813**, 1409–1414 (2006). Private communication (2006)
48. W. Gerlach, O. Stern, Das magnetische Moment des Silberatoms. Zeitschrift für Physik **9**, 353–355 (1922)
49. L. Hardy, Phys. Lett. A **167**, 11–19 (1992)
50. L. Hardy, Phys. Lett. A **175**, 259–260 (1993)
51. David Bohm, A suggested interpretation of the quantum theory in terms of 'hidden variables' I. Phys. Rev. **85**, 166–179 (1952)
52. R. Clifton, P. Neimann, Phys. Lett. A **166**, 177–184 (1992)
53. C. Pagonis, Phys. Lett. A **169**, 219–221 (1992)
54. R.B. Griffith, Phys. Lett. A **178**, 17 (1993)
55. C. Dewdny, L. Hardy, E.J. Squires, Phys. Lett. A **184**, 6–11 (1993)
56. A.C. Elitzur, S. Dolev, Multiple interaction-free measurements as a challenge to the transactional interpretation of quantum mechanics, ed. by D. Sheehan, in *Frontiers of Time: Retrocausation - Experiment and Theory*, vol. 863, AIP Conference Proceedings (2006), pp. 27–44
57. J.G. Cramer, Found. Phys. Lett. **19**, 63–73 (2006)
58. A.C. Elitzur, S. Dolev, A. Zeilinger, 'Time-reversed EPR and the choice of histories in quantum mechanics, in *Proceedings of XXII Solvay Conference in Physics*, World Scientific, New York (2002). arXiv:0205182 [quant-ph]
59. A.J. Leggett, Supp. Prog. Theor. Phys. **69**, 80 (1980)
60. A.J. Leggett, Anupam Garg, Quantum mechanics versus macroscopic realism: is the flux there when nobody looks? Phys. Rev. Lett. **54**, 857 (1985)
61. Anthony J. Leggett, Found. Phys. **33**, 1469 (2003)
62. S. Grblacher, T. Paterek, R. Kaltenbaek, C. Brukner, M. Zukowski, M. Aspelmeyer, A. Zeilinger, Nature **446**, 871–875 (2007), arXiv:0704.2529 [quant-ph]
63. W.H. Furry, Phys. Rev. **49**, 393 (1936a)
64. W.H. Furry, Phys. Rev. **49**, 476 (1936b)
65. M. Żukowski, A. Zeilinger, M.A. Horne, A.K. Ekert, Phys. Rev. Lett. **71**, 4287 (1993)
66. R. Kaltenbaek, R. Prevedel, M. Aspelmeyer, A. Zeilinger, High-fidelity entanglement swapping with fully independent sources. Phys. Rev. A **79**, 040302 (2009)
67. C.K. Hong, Z.Y. Ou, L. Mandel, Phys. Rev. Lett. **59**, 2044 (1987)
68. I. Marcikic, H. de Riedmatten, W. Tittel, H. Zbinden, M. Legré, N. Gisin, Phys. Rev. Lett. **93**, 180502 (2004)
69. H. Hübel, M.R. Vanner, T. Lederer, B. Blauensteiner, T. Lorünser, A. Poppe, A. Zeilinger, Opt. Express **15**, 7853 (2007)
70. R. Ursin, F. Tiefenbacher, T. Schmitt-Manderbach, H. Weier, T. Scheidl, M. Lindenthal, B. Blauensteiner, T. Jennewein, J. Perdigues, P. Trojek, B. Ömer, M. Fürst, M. Meyenburg, J. Rarity, Z. Sodnik, C. Barbieri, H. Weinfurter, A. Zeilinger, Entanglement-based quantum communication over 144 km. Nat. Phys. **3**, 481–486 (2007)
71. J.D. Bancal, S. Pironio, A. Achin, Y.-C. Liang, V. Scarani, N. Gisin, Quantum non-locality based on finite speed causal influences leads to superluminal signaling. Nat. Phys. **8**, 867–870 (2012), arXiv:1110.3795 [quant-ph]
72. S.W. Hawking, Particle creation by black holes. Commun. Math. Phys. **43**, 199–220 (1975)
73. A. Almheiri, D. Marolf, J. Polchinski, J. Sully, Black holes: complementarity or firewalls? J. High Energy Phys. **2013**(2), 062 (2013)
74. J. Maldacena, L. Susskind, Cool horizons for entangled black holes. Fortsch. Phys. **61**, 781–811 (2013)

# Chapter 7
# Nonlocal Signaling?

Given that a measurement on one part of an extended quantum system can affect the outcomes of measurements performed in other distant parts of the system, the question that naturally arises is: *can this phenomenon be used for nonlocal communication between one observer and another?* Demonstration of such nonlocal quantum communication would be a truly game-changing discovery, because it would break all the rules of normal communication. No energy would pass between the send and receive stations; the acts of sending and receiving could occur in either time order and would depend only on the observer-chosen instants at which the measurements were made; there would be no definite signal-propagation speed, and messages could effectively be sent faster than light-speed, or "instantaneously" in any chosen reference frame, or even, in principle, backwards in time. The average member of the physics community, if he or she has any opinion about nonlocal communication at all, believes it to be impossible, in part because of its superluminal and retrocausal implications.

## 7.1 No-Signal Theorems

Over the years a number of authors have presented proofs, based on the standard quantum formalism, showing that nonlocal observer-to-observer communication is impossible [1–4]. They employ details of quantum mechanics and quantum field theory and show that in separated measurements involving entangled quantum systems, the quantum correlations will be preserved but there will be no effect apparent to an observer in one sub-system if the character of the measurement and observation are changed in the other sub-system. Thus, the standard quantum formalism implies that nonlocal signaling is impossible, and any hypothetical observation of nonlocal signaling would require some change in that formalism.



J.G. Cramer, *The Quantum Handshake*, DOI 10.1007/978-3-319-24642-0_7

However, there is another viewpoint on the issue of no-signaling. A number of authors [5–10], mainly philosophers of science, have pointed out that the "proofs" ruling out nonlocal signaling are in some sense tautological, because they assume that the measurement process and its associated Hamiltonian are local, thereby effectively building-in the final no-signal conclusion from the starting assumptions. To quote Peter Mittelstaedt [8]:

> It turns out that the (no-signaling) arguments … are merely plausible but not really stringent and convincing. This means that the question (of nonlocal signaling) is still open.

Moreover, one successful superluminal experiment would trump all of the theoretical impossibility proofs.

This criticism by philosophers is more a complaint about the quantum formalism itself than about the authors who have used it. The criticized tautological assumptions, in some sense, have been deliberately embedded in the current formalism of quantum mechanics and quantum field theory, placed there by its formulators who used no-signaling as a guide. Can the bias against nonlocal signaling be removed from the quantum formalism? To adopt less restrictive and prejudicial tenets that lead to the same predictions and observations would require modification of the standard formalism in ways that have not, to my knowledge, been proposed or implemented. In any case, it should not be surprising that application of the current quantum formalism, with its built-in biases, can be expected to "demonstrate" that nonlocal signaling is not possible.

Even if nonlocal signaling is incompatible with the current formalism of quantum mechanics, it is interesting to observe in specific seemingly paradoxical cases (see below) exactly how the potential nonlocal signal is blocked. Such understanding could suggest ways in which new experiments based on a generalized formalism might actually lead to nonlocal signaling, or at least push the envelope in that direction.

## 7.2 Nonlocal Signals and Special Relativity

It is sometimes asserted that nonlocal communication is clearly impossible because it would conflict with special relativity. This assertion is not correct. The prohibition of signals with superluminal speeds by Einstein's theory of special relativity is related to the fact that a condition of definite simultaneity between two separated space-time points is not Lorenz invariant. Assuming that some hypothetical superluminal signal could be used to establish a fixed simultaneity relation between two such space-time points, e.g., by clock synchronization, this would imply a preferred inertial frame and would be inconsistent with Lorenz invariance and special relativity. In other words, superluminal signaling would be inconsistent with the even-handed treatment of all inertial reference frames that is the basis of special relativity.

However, if a hypothetical nonlocal signal could be transmitted through measurements at separated locations performed on a pair of entangled photons, the signal would be "sent" at the time of the arrival of one photon at one location and "received" at the time of arrival of the other photon at the other location, both along Lorenz-invariant light-like world lines. By varying path lengths and time-delays to the two locations, these events could be made to occur in any order and time separation in any reference frame. Therefore, nonlocal signals (even superluminal and retrocausal ones) could *not* be used to establish a fixed simultaneity relation between two separated space-time points, because the sending and receiving of such signals do not have fixed time relations. Nonlocal quantum signaling, if it were to exist, would be completely compatible with special relativity. (However, it would probably *not* be compatible with macroscopic causality.)

## 7.3 Entanglement-Coherence Complementarity and Variable Entanglement

One plausible mechanism for blocking nonlocal signals is the known complementary relation between entanglement and coherence [11]. In a two photon system with perfect entanglement, the wave functions of the two photons will have no definite phase relation, i.e., no coherence. There is a "see-saw" relation between the squares of the quantities specifying entanglement ($\varepsilon$) and coherence ($\kappa$): $\varepsilon^2 + \kappa^2 = 1$. This means that 100 % entanglement implies no coherence and *vice versa*. Since some degree of *both* coherence and entanglement would be required for any potential nonlocal signal [12], it is particularly interesting to study systems in which the entanglement/coherence ratio is a parameter that can be varied and set at 71 % of each (see below). Since the no-signal theorems usually assume complete entanglement, there remains the question of whether they apply to the case of partial entanglement and partial coherence.

Fortunately, the effects of variable entanglement can be investigated experimentally because the Zeilinger Group [13], has developed a Sagnac-mode two-photon source in which the degree of entanglement between the photons can be varied depending on the variable $\alpha$, producing a two-particle wave function of the general form:

$$\begin{aligned}\Psi(\alpha) = &(|H_A\rangle\,|H_B\rangle + |V_A\rangle\,|V_B\rangle)\sin(\alpha)\sqrt{2}\\ &- i(|H_A\rangle\,|V_B\rangle - |V_A\rangle\,|H_B\rangle)\cos(\alpha)/\sqrt{2},\end{aligned} \tag{7.1}$$

where $\alpha$ is an angle by which a half-wave plate is rotated within the source. In other words, the source produces a linear combination of the two orthogonal Bell states of Eqs. 2.1 and 2.2. The degree of photon-pair entanglement from this source is adjustable. When $\alpha = 0$, the two-photon polarization entanglement is 100 % in a pure antisymmetic Bell state (see Eq. 2.2) with the wave function:

$$\Psi(0) = i(|H_A\rangle\,|V_B\rangle - |V_A\rangle\,|H_B\rangle)/\sqrt{2}. \tag{7.2}$$

When $\alpha = \pi/4$ the entanglement is 0 in a coherent non-entangled product state with:

$$\Psi(\pi/4) = [(|H_A\rangle - i\,|V_A\rangle) \times (|H_B\rangle + i\,|V_B\rangle)]/2. \tag{7.3}$$

When $\alpha = \pi/2$, the two-photon polarization entanglement is 100 % in a pure symmetric Bell state (see Eq. 2.1) with the wave function:

$$\Psi(\pi/2) = (|H_A\rangle\,|H_B\rangle + |V_A\rangle\,|V_B\rangle)/\sqrt{2}. \tag{7.4}$$

As an intermediate case, when $\alpha = \pi/8$ the source will produce photon pairs with 71 % entanglement and 71 % coherence, with a 1:1 linear combination of Eqs. 7.2 and 7.3.

## 7.4 A Polarization-Entangled EPR Experiment with Variable Entanglement

To consider hypothetical nonlocal signaling with variable entanglement, let us begin by examining a fairly simple EPR experiment exhibiting nonlocality. Following Bell [14], a number of experimental EPR tests [15–17] have exploited the correlations of polarization-entangled systems that arise from angular momentum conservation. Their results, to an accuracy of many standard deviations, are consistent with the predictions of standard quantum mechanics and can be interpreted as falsifying many local hidden-variable alternatives to quantum mechanics.

A modern version of this type of EPR experiment, one in which the entanglement/coherence ratio is an adjustable parameter using the source described in Sect. 7.3 above, is shown in Fig. 7.1. We note that while there are many analyses of such experiments in the literature, to our knowledge there is no previous analysis for an EPR system with variable source entanglement [18].

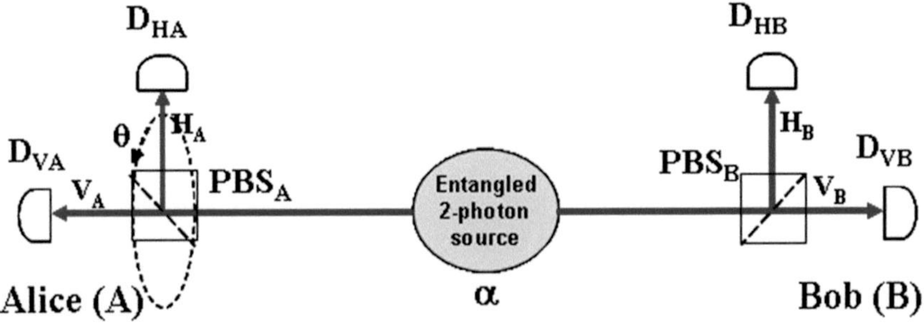


**Fig. 7.1** A two-photon 4-detector EPR experiment using linear polarization with variable entanglement. Here $\alpha$ is the parameter that varies the entanglement of photon pairs from the source

Two observers, Alice and Bob, operate polarimeters measuring the linear polarization (H or V) of individual photons and record photon detections. The H-V plane of Alice's polarimeter can be rotated through an angle $\theta$ with respect to the plane of Bob's polarimeter, so that the basis of her polarization measurements can be changed relative to Bob's. Appendix D.2 provides a detailed analysis of this experiment.

Now consider the question of whether, at any setting of $\alpha$, observer Alice by operating the left system and varying $\theta$ can send a nonlocal signal to observer Bob operating the right system. Some overall observer who is monitoring the coincidence counting rates $H_A H_B$, $V_A V_B$, $H_A V_B$, and $V_A H_B$ could produce a map of the correlations (see Fig. D.4 in Appendix D.2) and would have a clear indication of when $\theta$ was varied by Alice, in that the relative rates would change dramatically. However, observer Bob is isolated at the system on the right and is monitoring only the two singles counting rates $H_B \equiv H_A H_B + V_A H_B$ and $V_B \equiv H_A V_B + V_A V_B$.

Bob would observe the probabilities $P_{BH}(\alpha, \theta) = P_{HH}(\alpha, \theta) + P_{VH}(\alpha, \theta) = 1/2$ and $P_{BV}(\alpha, \theta) = P_{HV}(\alpha, \theta) + P_{VV}(\alpha, \theta) = 1/2$, both independent of the values of $\alpha$ and $\theta$. Thus, Bob would see only counts detected at random in one or the other of his detectors with a 50 % chance of each polarization, and his observed rates would not be affected by the setting of $\theta$. Alice's choice of her $\theta$ setting will alter the wave functions that arrive at Bob's detectors, but not in a way that permits signaling. Schrödinger called this effect "steering" the wave functions [19]. The late Heinz Pagels, in his book *The Cosmic Code* [20], examined in great detail the way in which the intrinsic randomness of quantum mechanics blocks any potential nonlocal signal in this type of polarization-based EPR experiment.

We emphasize the point that linear polarization is an interference effect of the photon's intrinsic circularly-polarized spin angular momentum $S = 1$, $S_z = \pm 1$ helicity eigenstates. As we will see below, the interference blocking observed here is an example of a "signal" interference pattern and an "anti-signal" interference pattern that mask any observable interference when they are added, even when entanglement and coherence are simultaneously present. This behavior is attributed to what has been called the complementarity of one-particle and two-particle interference [20].

## 7.5 A Path-Entangled EPR Experiment with Variable Entanglement*

Although the entanglement of linear polarization is a very convenient medium for EPR experiments and Bell-inequality tests, in many ways the alternative offered by path-entangled EPR experiments provides a richer venue. Perhaps the earliest example of a path-entangled EPR experiment is the 1995 "ghost interference" experiment of the Shih Group at University of Maryland Baltimore County [22] discussed in Sect. 6.16.1. Their experiment demonstrated that an interference pattern observed for one member of a pair of entangled photons could be "switched" off or on depending on whether the other photon of the pair went through one slit or both slits of a two-slit aperture.

Another path-entangled EPR experiment was the 1999 PhD thesis of Dr. Birgit Dopfer at the University of Innsbruck [23], discussed in Sect. 6.16.2. This experiment demonstrated that one could can make the interference pattern observed for one of a pair of entangled photons appear or disappear, depending on whether the location of the detector that detected the other member of the entangled pair was at or away from the focal point of a lens.

Examination of the ghost-interference and Dopfer experiments raises a very interesting question: Can the requirement of a coincidence between the entangled photons, used in both experiments, be removed while preserving the switchable interference pattern? The answer to this question is subtle. In principle, the two entangled photons are connected by nonlocality whether they are detected in coincidence or not, so the coincidence may perhaps be removable. However, in both experiments the authors reported that no two-slit interference distribution was observed when the coincidence requirement was removed. These considerations lead to a new quantum mechanical paradox: they suggest [24] that if the coincidence requirement could be relaxed, nonlocal observer-to-observer signals might be transmitted by controlling the presence or absence of an interference pattern, essentially by forcing wave-like or particle-like behavior on both members of an entangled photon pair. And since a two-slit interference pattern only becomes visible when many single-photon events are observed (see Fig. 6.2), it might be argued that using a two-slit interference pattern as a signal evades the no-signal prohibition because the latter only applies to single quantum events, which individually carry no information. (As we will see, this argument is incorrect.)

From the point of view of moving to a path-entanglement situation in which the coincidence requirement could be relaxed, the problem with both of the experiments discussed above is their use of a two-slit system that blocks and absorbs most of the photons from the nonlinear crystal that illuminate the slit system. The down-conversion process is intrinsically very inefficient ($\sim$1 photon pair per $10^8$ pump photons), so there are no photons to waste. An additional complication is that most detectors capable of detecting individual photons are intrinsically noisy and somewhat inefficient. For these reasons, there is a large advantage in using *all* of the available entangled-photon pairs in any contemplated path-interference test of nonlocal communication.

Figure 7.2 shows a path-entangled experimental EPR test using two Mach–Zehnder interferometers [25, 26] that have been modified to convert polarization entanglement to path entanglement [27]. This type of system was originally developed by the Zeilinger Group at the Institute for Quantum Optics and Quantum Information, Vienna [28]. Here, the interferometers are a variant of the basic Mach–Zehnder design that uses an initial polarizing beam splitter ($PBS_{A,B}$) that directs the vertical ($v$) and horizontal ($h$) linear polarizations to different paths and then converts horizontal to vertical polarization on the upper path with a half-wave plate ($HW_{A,B}$). This has the effect of converting polarization entanglement from the source to path entanglement and then placing waves on both paths in the same polarization state, so that they can interfere. Again observers Alice and Bob operate the interferometers and count and record individual photon detections. A phase shift element ($\phi_{A,B}$)

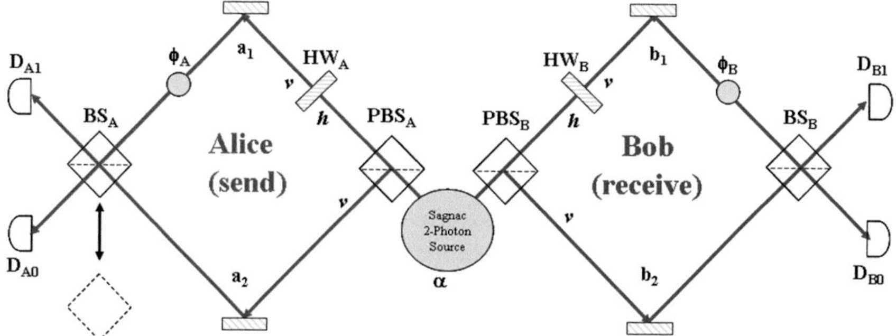


**Fig. 7.2** A 4-detector path-entangled dual-interferometer EPR experiment with variable entanglement

allows the observers to alter the phase of waves on the upper paths. A final 50:50 polarization-insensitive beam splitter $BS_{A,B}$ at the cross point of the beams re-mixes them and sends them to single-photon detectors $D_{A,B0}$ and $D_{A,B1}$.

As in the polarization-entangled EPR example of Sect. 7.3, the source of photons is taken to be the Sagnac entangled two-photon source developed by the Zeilinger Group [13], in which the degree of entanglement depends on the value of $\alpha$ according to Eqs. 7.1–7.3.

Alice's last beam-splitter ($BS_A$) is removable, as indicated by the dashed outline in Fig. 7.2. When $BS_A$ is in place, the two left paths are remixed, the left-going photons exhibit the wave-like behavior of being on both paths, and two-path overlap and Mach–Zehnder interference will be present. When $BS_A$ is removed, path detection occurs, the left-going photons exhibit the particle-like behavior of being on a path uniquely ending at detector $D_{A0}$ or at detector $D_{A1}$, so that Alice's measurements provide which-way information about both photons. Bob's last beam splitter ($BS_B$) remains in place and, in the absence of which-way information, should exhibit Mach–Zehnder interference.

This experiment is thus the equivalent of the ghost-interference experiment and the Dopfer experiment described above, in that it embodies entangled paths and two-path interference. However, it improves on those experiments by using all of the available entangled photons and by employing a source that has a variable entanglement that depends on $\alpha$.

It has been argued [12, 24] that this situation presents a nonlocal signaling paradox, in that Alice, by choosing whether $BS_A$ is in or out, can cause the Mach–Zehnder interference effect to be present or absent in Bob's detectors. In particular, with $BS_A$ out we expect particle-like behavior, and Bob should observe equal counting rates in $D_{B1}$ and $D_{B0}$. With $BS_A$ in we expect wave-like behavior, and Bob, for the proper choice of $\phi_B$, should observe all counts in $D_{B1}$ and no counts in $D_{B0}$ due to Mach–Zehnder interference. It was further argued [12] that possibly the nonlocal signal might be suppressed by the complementarity of entanglement and coherence [11], but that by arranging for 71 % entanglement and 71 % coherence (i.e., $\alpha = \pi/8$ for the Sagnac source), a nonlocal signal might be permitted.

Appendix D.3 provides a detailed analysis of this experiment. The conclusion is that, for any value of $\alpha$, no nonlocal signal can be sent by inserting and removing $BS_A$ or by varying phase $\phi_A$. We have also found (not shown here) that even when the left-going photons from the source are intercepted *before* entering Alice's interferometer with a black absorber, Bob will still observe the same singles counting rates given by Eqs. D.37 and D.38. As in the polarization-entangled EPR case, the interference blocking observed is an example of a "signal" interference pattern and an "anti-signal" interference pattern that may individually vary dramatically with Alice's choice of measurement due to Schrödinger steering, but that will mask any observable interference when they are added, even when entanglement and coherence are simultaneously present. This behavior is again attributed to what has been called the complementarity of one-particle and two-particle interference [21].

## 7.6 A Wedge-Modified Path-Entangled EPR Experiment with Variable Entanglement

A possible reason that the above attempts at nonlocal communication have failed is that the left-going photons are directed to both of Alice's detectors. The two detectors measure complementary interference profiles, so that when these profiles are added the potential nonlocal signal is erased. Suppose that instead we direct all of the left-going photons on both paths to a single detector, where they should have only one interference profile. Could this change permit nonlocal signaling? To investigate this question we have analyzed the experiment shown in Fig. 7.3.

Here, we have replaced Alice's last beam splitter and detectors with a somewhat unorthodox optical device, a 45° wedge mirror $W_A$ that directs the left-going photons on paths $a_1$ and $a_2$ to a single detector $D_A$. We assume that the angles of Alice's mirrors are tweaked slightly so that the two beams have a maximum overlap at $D_A$ and that $W_A$ is positioned so that it reflects most of the two beams, except for their extreme Gaussian tails ($\sim 10\sigma$). Also, a removable beam stop has been placed in the path of the left-going photons near the source. As stated above, when the left-going photons

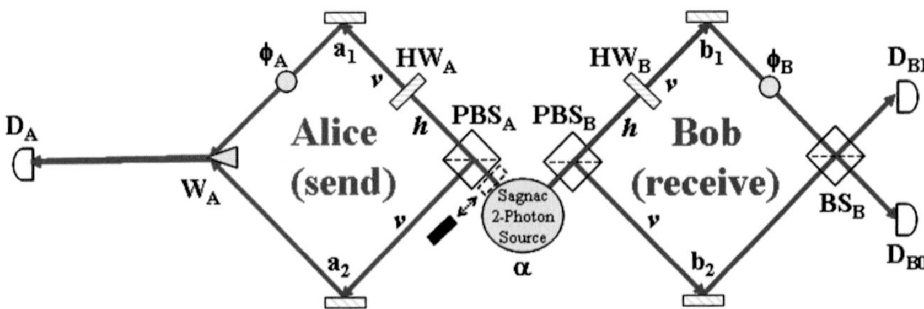


**Fig. 7.3** A 3-detector wedge modification of the path-entangled dual-interferometer EPR experiment with variable entanglement

from the source are intercepted by such a beam stop, the non-coincident singles probabilities for Bob's detectors will be given by Eqs. D.37 and D.38. We wish to investigate the question of whether Bob will observe any change in the counting rates of his detectors that depends on whether the beam stop is in or out.

Naively it might appear that the new configuration would produce a large change in Bob's counting rates, because Alice could choose a phase $\phi_A$ for which the left-going wave components arriving at $D_A$ would interfere destructively and vanish or would interfere constructively and produce a maximum. Arguments along these lines have been advanced by Anwar Shiekh [29] to justify a clever (but flawed) one-photon faster-than-light communication scheme. However, such expectations cannot be true, because they would violate quantum unitarity and the requirement that any left-going photon must be detected *somewhere* with 100 % probability. Unitarity (or equivalently, energy conservation) requires that any wave-mixing device that produces destructive interference in some locations must produce a precisely equal amount of constructive interference in other locations. The 45° wedge beam-combiner is no exception.

The flaw in such cancellation arguments is that in the previous examples we have always dealt with configurations in which only a single spatial mode of the photon is present. In such cases, superposition can be used without considering wave trajectories, since the wave front for any given path arrives at a detector with a constant overall phase. In the present configuration, the spatial profiles of the waves on Alice's two paths are truncated at the apex of the wedge mirror, producing non-Gaussian spatial modes, and also must propagate in slightly different directions in order to overlap at the detector so they are definitely in different spatial modes. Therefore, the phase of arriving waves is not constant and will depend on the location on the detector face. Consequently, simple one-mode position-independent superposition cannot be used.

Instead, in order to calculate the differential probability of detection at a specific location on the face of detector $D_A$, one must propagate the waves from the wedge to the detector by doing a path integral of Huygens wavelets originating across the effective aperture of the wedge. To get the overall detection probability, one must then integrate over locations on the detector face. And since there are two quantum-distinguishable amplitudes arriving at the detector face, these must be converted to probabilities separately and then added.

The analysis of the wedge system is therefore much more challenging than those of the previous examples. While analytic expressions can be obtained for the differential probability of two-particle detection with one of Bob's detectors and at some specific lateral position on $D_A$, the integration of that differential probability, a highly oscillatory function, over the face of $D_A$ cannot be done analytically. Thus the analysis cannot produce equations predicting Bob's singles counts that can be directly compared with Eqs. D.37 and D.38 for the signal test. Instead one must subtract the results of numerical integration from evaluations of Eqs. D.37 and D.38 using the same values for $\alpha$, $\phi_A$, and $\phi_B$ used in the numerical integration, and observe how close to zero is the calculated difference (which represents the potential nonlocal signal).

Appendix D.4 provides a detailed analysis of this experiment. Our conclusion, based on the standard formalism of quantum mechanics as applied to these *gedanken-experiments*, is that no nonlocal signal can be transmitted from Alice to Bob by varying Alice's configuration in any of the ways discussed here. In all of the cases studied, there are two quantum-distinguishable modes of entangled photon-pair behavior that each contain a "switch-able" interference pattern, but when these two modes are superimposed, the two interference patterns always complement each other and together become invisible. This is the mechanism by which the formalism of quantum mechanics blocks nonlocal signaling. In the context of the standard quantum formalism, Nature appears to be well protected from the possibility of nonlocal signaling.

## 7.7 A Transactional Analysis of the Complementarity of One- and Two-Particle Interference

The transactions that form in the experiment shown in Fig. 7.2 are so-called "V" transactions, as discussed for the Freedman–Clauser EPR experiment [15] and shown in Fig. 6.9. These have a pair of offer waves of the entangled photon pair starting at the source and traveling to one of the "A" detectors on the left and one of the "B" detectors on the right. The arriving offer waves generate confirmation waves that start at these detectors and travel back to the source. There are three vertices in such a transaction ($D_{Ai}$, the source, and $D_{Bi}$), where $i = 0$ or 1 and at each of these vertices the offer and confirmation paths of the waves must match for the transaction to go on to completion, resulting in the transfer of photon energy, momentum, and spin from the source to the detectors.

However, there is another way to look at this type of transaction. Because the confirmation wave from one of Bob's detectors to the source connects with an offer wave from the source to one of Alice's detectors (and similarly in the other direction), the "V" transaction is equivalent to a transaction directly between one of Bob's detectors and one of Alice's detectors, and we can analyze the system from that point of view. Further, there will be a Wheeler–Feynman 180° phase change between the connecting offer and confirmation waves across the source, but because it affects both paths equally, it can be ignored.

Now let us focus on the transactions that occur at Bob's upper detector $D_{B1}$, using this way of viewing "V" transactions. First, consider the transactions connecting Alice's upper detector $D_{A1}$ with Bob's upper detector $D_{B1}$, as shown in Fig. 7.4. Because of the two-photon entanglement, an offer/advanced wave connection from $D_{B1}$ to $D_{A1}$ and back must either take the upper path (red/solid) or the lower path (blue/dashed). With $\alpha = \pi/2$ there will be no change in phase between the connecting offer and confirmation waves across the source. When the phase shifters $\phi_A$ and $\phi_B$ are set to zero phase change, the only phase changes present will be produced by the four reflections on the upper (red/solid) path or the four reflections on the

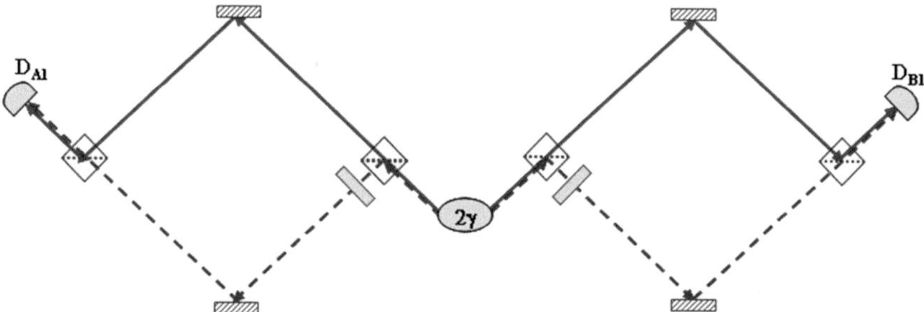


**Fig. 7.4** A transaction linking $D_{A1}$, the 2-photon source ($2\gamma$), and $D_{B1}$. The entanglement is such that one dual offer wave travels only on the upper paths (*red/solid*) and another travels only on the lower paths (*blue/dotted*). These amplitudes interfere constructively at the three vertices, and photons will be detected in coincidence at $D_{A1}$ and $D_{B1}$

lower (blue/dashed) path, with each 90° reflection introducing a phase shift of 90°. Therefore, the waves arriving at $D_{A1}$ on the two paths will be in phase and will reinforce, producing a complete two-photon transaction between $D_{A1}$ and $D_{B1}$, and each will detect a photon.

Next, consider the transactions connecting $D_{A0}$ and $D_{B1}$ shown in Fig. 7.5. An offer/advanced wave connection from $D_{B1}$ to $D_{A0}$ and back must either take the upper path (red/solid) or the lower path (blue/dashed). Now there will be three reflections on the upper (red/solid) path and five reflections on the lower (blue/dashed) path. Therefore, the waves arriving at $D_{A0}$ on the two paths will be 180° out phase, will destructively interfere, and will cancel. Therefore, no overall two-photon transaction can form between $D_{A0}$ and $D_{B1}$, and no coincident photons will be detected by this detector combination. For this reason, for photon detection at $D_{B1}$, all of the coincident photons will be detected at $D_{A1}$ and none will be detected at $D_{A0}$.

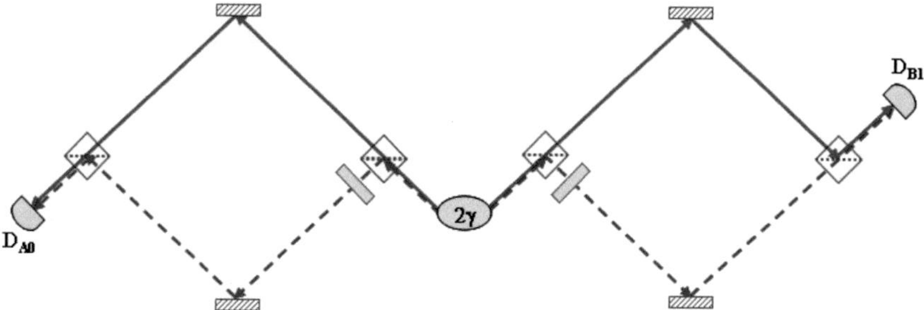


**Fig. 7.5** A transaction linking $D_{A0}$, the 2-photon source ($2\gamma$), and $D_{B1}$. The entanglement is such that one dual offer wave travels only on the upper paths (*red/solid*) and another travels only on the lower paths (*blue/dotted*). These amplitudes interfere destructively at the three vertices, and no photons will be detected at $D_{A0}$ and $D_{B1}$

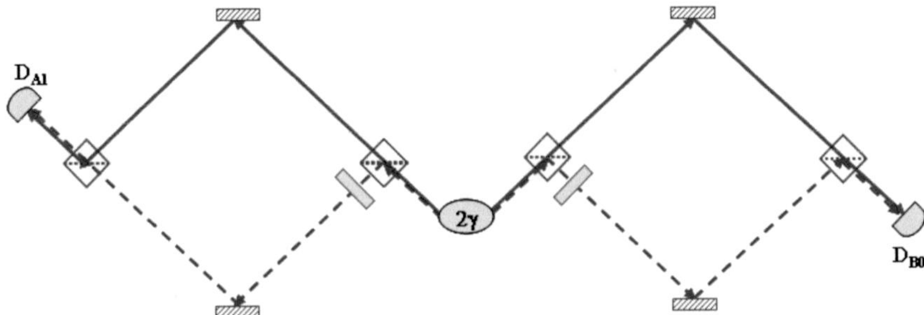


**Fig. 7.6** A transaction linking $D_{A1}$, the 2-photon source ($2\gamma$), and $D_{B0}$. The entanglement is such that one dual offer wave travels only on the upper paths (*red/solid*) and another travels only on the lower paths (*blue/dotted*). These amplitudes interfere destructively at the three vertices, and no photons will be detected at $D_{A1}$ and $D_{B0}$

What about Bob's lower detector? The transactions connecting Alice's detector $D_{A1}$ with $D_{B0}$ are shown in Fig. 7.6. Again, an offer/advanced wave connection from $D_{B0}$ to $D_{A1}$ and back must either take the upper path (red/solid) or the lower path (blue/dashed). Again there will be three reflections on the upper (red/solid) path and five reflections on the lower (blue/dashed) path. Therefore, the waves arriving at $D_{A1}$ on the two paths will be 180° out phase, will destructively interfere, and will cancel. Again, no overall two-photon transaction can form between $D_{A1}$ and $D_{B0}$, and no coincident photons will be detected by this detector combination.

Finally, consider the transactions connecting $D_{A0}$ and $D_{B0}$ shown in Fig. 7.7. An offer/advanced wave connection from $D_{B0}$ to $D_{A0}$ and back must again take either the upper path (red/solid) or the lower path (blue/dashed). Now there will be two reflections on the upper (red/solid) path and six reflections on the lower (blue/dashed) path. Therefore, the waves arriving at $D_{A10}$ on the two paths will be in phase, will constructively interfere, and will reinforce, producing a complete overall two-photon

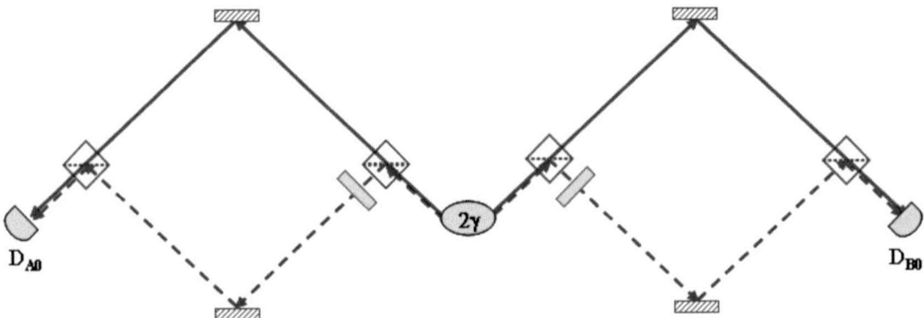


**Fig. 7.7** A transaction linking $D_{A0}$, the 2-photon source ($2\gamma$), and $D_{B0}$. The entanglement is such that one dual offer wave travels only on the upper paths (*red*/solid) and another travels only on the lower paths (*blue/dotted*). These amplitudes interfere constructively at the three vertices, and photons will be detected in coincidence at $D_{A0}$ and $D_{B0}$

transaction between $D_{A0}$ and $D_{B0}$, and each will detect a photon. For detection at $D_{B0}$, all of the coincident photons will be detected at $D_{A0}$ and none will be detected at $D_{A1}$.

## 7.8 Singles Detection and the Absence of 1-Particle Interference

Because the system in Fig. 7.1 used two modified Mach–Zehnder interferometers [25, 26], the interference effect shows up as channeling all photons to one detector and none to the other, following the final 50:50 beam splitter. We have seen two-particle interference effects appearing in the coincident detections discussed in Sect. 7.4 above.

Now we will consider one-particle interference in Bob's detectors. Because a photon detected in one of Bob's detectors must be in coincidence with one or the other of Alice's detectors, and because these two detection combinations are quantum-mechanically distinguishable and incoherent and do not interfere, we must add the two coincident probabilities (not amplitudes) involving a given detector of Bob's to obtain the "singles" detection probability for that detector. As we have seen above, when $\alpha = \pi/2$ and $\phi_A = \phi_B = 0$, each of Bob's detectors has a 50 % probability of being in coincidence with one of Alice's detectors and a 0 % probability of being in coincidence with the other. Thus, Bob observes a random 50:50 split between the photons going to his two detectors, exactly as if there was no interference at all. We note from Sect. 7.4 that this will also be the case for any other settings of $\alpha$ and $\phi_A$.

This is an example of the operation of the complementarity between two-particle and one-particle interference [21]: opposite two-particle interference patterns, which must be added to obtain the singles probability, erase any sign of a one-particle interference pattern. This appears to be a general characteristic of the formalism of quantum mechanics. As seen in our inquiry into nonlocal signaling discussed above, it prevents the occurrence of potentially retrocausal signals. In the arrangement shown, entanglement insures that Bob receives two quantum-distinguishable streams of waves from Alice that have opposite phases, produce opposite interference patterns, and always, for any setting of $\phi_A$, will add up to no interference pattern at all.

The underlying cause of this suppression of one-particle interference is that geometry causes a 180° switch in the relative phase between the two waves arriving at the two left detectors, which reverses the interference pattern. This correlation of deflection angle with phase shift is an inescapable feature of quantum optics, and it provides the mechanism that "builds in" the observed interference complementarity.

## 7.9 Entangled Paths and Hidden Signals: A Proof

Another approach to the question of whether path-entangled EPR experiments can be used to transmit nonlocal signals using switchable interference patterns has been recently addressed in a very general way by Nick Herbert [30]. He derived a simpler version of the proof below, which uses the properties of orthogonal basis states. We reproduce and generalize the proof here with his permission.

Suppose that we have a source $S$ of entangled photon pairs, delivering entangled photons to experimenters Alice and Bob in a dual interferometer setup like those described above. The variable-entanglement source output is that given by Eq. 7.1, but with paths substituted for polarizations. The source wave function $\Psi$ is therefore a linear combination of symmetric and antisymmetric Bell-states. It has the form:

$$\Psi(\alpha) = (|1\rangle_A\,|1\rangle_B + |2\rangle_A\,|2\rangle_B)\sin(\alpha)\sqrt{2} \\ - i(|1\rangle_A\,|2\rangle_B - |2\rangle_A\,|1\rangle_B)\cos(\alpha)/\sqrt{2}, \tag{7.5}$$

where $|\,1\rangle_A$ is the wave function for waves on Alice's path $a_1$, $|\,2\rangle_A$ is the wave function for waves on Alice's path $a_2$, $|\,1\rangle_B$ is the wave function for waves on Bob's path $b_1$, and $|\,2\rangle_B$ is the wave function for waves on Bob's path $b_2$.

These path-based wave function components comprise an orthogonal basis set much like the $H/V$ basis discussed in Sect. 6.7 above. Now let us define wave functions using a new orthogonal set, the interference basis, by combining the $A$ or $B$ paths, as might be done by the action of splitters like the ones in Fig. 7.2.

$$|\,A\rangle_+ \equiv \frac{1}{\sqrt{2}}(|\,1\rangle_A + |\,2\rangle_A) \tag{7.6}$$

$$|\,A\rangle_- \equiv \frac{1}{\sqrt{2}}(|\,1\rangle_A - |\,2\rangle_A) \tag{7.7}$$

$$|\,B\rangle_+ \equiv \frac{1}{\sqrt{2}}(|\,1\rangle_B + |\,2\rangle_B) \tag{7.8}$$

$$|\,B\rangle_- \equiv \frac{1}{\sqrt{2}}(|\,1\rangle_B - |\,2\rangle_B) \tag{7.9}$$

The inverse relations that connect to back to the path basis are:

$$|\,1\rangle_A = \frac{1}{\sqrt{2}}(|\,A\rangle_+ + |\,A\rangle_-) \tag{7.10}$$

$$|\,2\rangle_A = \frac{1}{\sqrt{2}}(|\,A\rangle_+ - |\,A\rangle_-) \tag{7.11}$$

$$|\,1\rangle_B = \frac{1}{\sqrt{2}}(|\,B\rangle_+ + |\,B\rangle_-) \tag{7.12}$$

$$|\,2\rangle_B = \frac{1}{\sqrt{2}}(|\,B\rangle_+ - |\,B\rangle_-) \tag{7.13}$$

Note that both of these bases are sets of four orthogonal states.

Here, $| A\rangle_+$ represents the situation in which the waves on Alice's two paths *interfere*, $| A\rangle_-$ represents the situation in which the waves on Alice's two paths *anti-interfere*, $| B\rangle_+$ represents the situation in which the waves on Bob's two paths *interfere*, and $| B\rangle_-$ represents the situation in which the waves on Bob's two paths *anti-interfere*. The terms "interfere" and "anti-interfere" differ in that in anti-interference the relative phase of the two waves is increased by an additional 180°, thereby inverting any interference pattern by reversing regions of constructive and destructive interference. Now we substitute Eqs. 7.10–7.13 into Eq. 7.5 to obtain the variable-entanglement source wave function in terms of the new interference basis:

$$\begin{aligned}\Psi(\alpha) = (|A\rangle_+ \ |B\rangle_+ + \ |A\rangle_- \ |B\rangle_-) \sin(\alpha)\sqrt{2} \\ - i(|A\rangle_+ \ |B\rangle_- - \ |A\rangle_- \ |B\rangle_+) \cos(\alpha)/\sqrt{2},\end{aligned} \tag{7.14}$$

Equation 7.14 represents a linear combination of symmetric and antisymmetric Bell-state entanglement of two complementary interference patterns. For any measurement, i.e., any formation of a three-vertex transaction connecting the source to one of Alice's detectors and one of Bob's, the first term will collapse into either $| A\rangle_+ \ | B\rangle_+$ or $| A\rangle_- \ | B\rangle_-$ and the second term will collapse into either $| A\rangle_+ \ | B\rangle_-$ or $| A\rangle_- \ | B\rangle_+$ . If $\alpha = \pi/2$, in coincidence Alice and Bob will either measure interference together or they will measure anti-interference together. If $\alpha = 0$ they will measure correlated interference/anti-interference combinations.

But when Bob looks in singles for any value of $\alpha$, he will receive an incoherent 1:1 superposition of $|| B\rangle_+|^2$ and $|| B\rangle_-|^2$, which leads to *no interference* at Bob's detectors, independent of whether they are combined by a beam-splitter, a double slit, or a wedge. Similarly, in singles, Alice will receive an incoherent 1:1 superposition of $|| A\rangle_+|^2$ and $|| A\rangle_-|^2$, which leads to *no interference* at Alice's detectors, independent of how they are combined. Thus, viewing a variable-entanglement dual interferometer experiment through the lens of the interference basis provides a very general proof of the impossibility of nonlocal signaling using path entanglement and interference switching.

## 7.10 Conclusions about Nonlocal Signals

We conclude, in the context of standard quantum mechanics, that a nonlocal signal cannot be achieved using any of the configurations considered. Switchable interference patterns, which seem to offer the possibility of signaling, are blocked by accompanying anti-interference patterns. Arranging for partial entanglement and partial coherence does not alter this situation.

There perhaps remains a dim possibility that if the quantum formalism were generalized to eliminate the built-in bias against nonlocal signals, a path to nonlocal signaling might be indicated by the new formalism. However, even if such a formalism was available, there is no guarantee that it would change the situation discussed above. The proof presented in Sect. 7.9 seems to demonstrate that signal blocking is a general consequence of orthogonal basis transformation.

## References


1. P.H. Eberhard, Nuovo Cimento B **38**, 75 (1977)
2. P.H. Eberhard, Nuovo Cimento B **46**, 392 (1978)
3. G.C. Ghirardi, A. Rimini, T. Weber, Lett. Nuovo Cimento **27**, 293 (1980)
4. U. Yurtsever, G. Hockney, Class. Quantum Gravity **22**, 295–312 (2005), arXiv:0409112 [gr-qc]
5. P.J. Bussey, Phys. Lett. A **123**, 1–3 (1987)
6. K.A. Peacock, Phys. Rev. Lett. **69**, 2733 (1992)
7. J.B. Kennedy, Philos. Sci. **62**, 543–560 (1995)
8. P. Mittelstädt, Ann. Phys. **7**, 710–715 (1998)
9. K.A. Peacock, B.S. Hepburn, in *Proceedings of the Meeting of the Society of Exact Philosophy* (1999). arXiv:quant-ph/9906036
10. S. Weinstein, Synthese **148**, 381–399 (2006)
11. A.F. Abouraddy, M.B. Nasr, B.E.A. Saleh, A.V. Sergienko, M.C. Teich, Phys. Rev. A **63**, 063803 (2001)
12. J.G. Cramer, Chapter 16 of, in *Frontiers of Propulsion Science*, ed. by M.G. Millis, E.W. Davis (American Institute of Aeronautics and Astronautics, Virginia, 2009). ISBN-10: 1-56347-956-7, ISBN-13: 978-1-56347-956-4
13. A. Fedrizzi, T. Herbst, A. Poppe, T. Jennewein, A. Zeilinger, Opt. Express **15**, 15377–15386 (2007)
14. J.S. Bell, Rev. Mod. Phys. **38**, 447 (1966)
15. S.J. Freedman, J.F. Clauser, Phys. Rev. Lett. **28**, 938 (1972)
16. A. Aspect, J. Dalibard, G. Roger, Phys. Rev. Lett. **49**, 91 (1982)
17. A. Aspect, J. Dalibard, G. Roger, Phys. Rev. Lett. **49**, 1804 (1982)
18. J.G. Cramer, N. Herbert, an inquiry into the possibility of nonlocal quantum communication, (submitted to Found. Phys.). arXiv:1409.5098 [quant-ph]
19. E. Schrödinger, Proc. Camb. Phil. Soc. **31**, 555–563 (1935)
20. Heinz Pagels, *The Cosmic Code* (Simon and Schuster, New York, 1982)
21. G. Jaeger, M.A. Horne, A. Shimony, Phys. Rev. A **48**, 1023–1027 (1993)
22. D.V. Strekalov, A.V. Sergienko, D.N. Klyshko, Y.H. Shih, Phys. Rev. Lett. **74**, 3600–3603 (1995)
23. B. Dopfer, Ph.D. thesis, University of Innsbruck (1998, unpublished); A. Zeilinger, Rev. Mod. Phys. **71**, S288–S297 (1999)
24. R. Jensen, in *Proceedings of STAIF 2006, AIP Conference Proceedings*, vol. 813, pp. 1409–1414 (2006) and private communication (2006)
25. L. Mach, Z. Instrumentenkunde **12**, 89 (1892)
26. L. Zehnder, Z. Instrumentenkunde **11**, 275 (1891)
27. M. Żukowski, J. Pykacz, Phys. Lett. A **127**, 1–4 (1988)
28. A. Fedrizzi, R. Lapkiewicz, X.-S. Ma, T. Paterek, T. Jennewein, A. Zeilinger, Demonstration of complementarity between one- and two-particle interference (October 21, 2008, unpublished preprint)
29. A.Y. Shiekh, Electr. J. Theor. Phys. **19**, 43 (2008)
30. N. Herbert, Private communication, used with permission (2015)

# Chapter 8
# Quantum Communication, Encryption, Teleportation, and Computing

## 8.1 Quantum Encryption and Communication and the TI

Quantum communication, i.e., the transmission of information using the properties of entangled photon pairs over significant distances, is rather difficult, because of the unavoidable absorption of photons in fiber-optics transmission lines and the intrinsic problem of efficient low-noise detection of single photons in the visible and infrared regions of the optical spectrum. Nevertheless, large quantum communication projects are under way in China and the USA, with a 2,000 km link planned between Beijing and Shanghai and a 650 km connection to Washington, D.C. planned by Battelle.

The attraction of such links is guaranteed security, which is related to the concepts of quantum encryption and key transmission. Basically, a quantum communication link would be untappable, because any intrusion into the link for surveillance would collapse the entangled photon states, destroying the message and leading to immediate detection of intrusion. That is apparently worth a great deal to governments and to certain commercial activities. Typically, the quantum link is used to super-securely transmit a very large encryption key, and this key, which is changed as often as possible, is used to decode encrypted messages sent by conventional transmission lines. There are also schemes for doubling the information content of such quantum-entangled transmissions by taking advantage of information stored in the entanglement.

The basic transmission problem is that for transmission distances greater than about 100 km, the probability of photon absorption in the fiber dominates the transmission and blocks the signal. Therefore, each 100 km or so a "quantum repeater" needs to be inserted to pass the signal along. The technique of entanglement swapping discussed in Sect. 6.22 provides a mechanism for constructing such repeaters using synchronized optical pumps. The Beijing-Shanghai quantum network plans to use 32 quantum repeaters in their system.



J.G. Cramer, *The Quantum Handshake*, DOI 10.1007/978-3-319-24642-0_8

As discussed in Sect. 6.22, the Transactional Interpretation description of advanced/retarded handshakes in the transmission process provides a readily understandable account of how the quantum entanglement is used for the transmission of information.

## 8.2 Quantum Teleportation and the TI

Teleportation is familiar idea in science fiction. It permeated the SF literature of the Golden Age, providing the basis for SF classics like A. E. Van Vogt's *World of Null-A* (1945), Alfred Bester's *The Stars My Destination* (1956), Algis Budrys' *Rogue Moon* (1960), and many others. The Star Trek TV series' transporter has also "beamed" the concept of teleportation into our pop culture. But modern hard-SF has largely abandoned teleportation as a concept that has more to do with fantasy and parapsychology than with real science.

Imagine my surprise then, to discover in 1993 an article [1] on the subject of teleportation in *Physical Review Letters*. The article, by an international collaboration of distinguished physicists hereafter designated by the initials BBCJPW, does not provide a plan for constructing a Star Trek transporter. Instead it describes an in-principle procedure for copying and transporting a pure quantum state from one location and one observer to another by a process that the authors characterize as teleportation. The BBCJPW scheme exploits some of the peculiarities of quantum physics, entanglement, and nonlocality, and it reveals some of the rules of the game in that exotic domain. Let's first discuss these rules.

First, we will focus on the quantum state vector or wave function. Any quantum system, for example an electron with the characteristics of position, energy, momentum, and a spin vector pointing in some direction, is completely described by the state vector, as denoted by the Dirac *ket* symbol $|\ y\rangle$. Anything that is knowable about the electron is mathematically encoded within $|\ y\rangle$. An essential rule of the quantum world is that the state vector can never be completely known because no measurement can determine it completely (except in the special case that it has been prepared in some particular state or some member of a known "basis" group of states in advance). In general, the quantum state coded within $|\ y\rangle$ can only be "glimpsed" by a measurement of one of the properties of the quantum system. As discussed in Sect. 2.4, in the act of pinning down one particular property of $|\ y\rangle$, the measurement destroys any opportunity to determine some of the other complementary properties of the quantum state. The quantum state can be preserved unchanged only by refraining from making any measurements of its properties. This frustrating aspect of quantum mechanics is the essence of Heisenberg's uncertainty principle.

A second ground rule of quantum mechanics is that a pair of spatially separated quantum sub-systems that are parts of an overall quantum system can be entangled, as discussed in Chap. 3. A measurement on one of the entangled sub-systems not only forces it into a particular state but also, across space-time and even backwards in time, forces the sub-system with which it is entangled into a corresponding state. For example, measuring the polarization of one of a pair of entangled photons precipitates the other photon, which may be light years away, into the same state of polarization as that which was measured for its entangled twin.

The state vector, however, does not have to describe a microscopic system like a photon or an electron. It can describe large collections of atoms: chemical compounds, human beings, planets, stars, galaxies. There have even been recent papers in quantum cosmology by Stephen Hawking and others in which the properties of the state vector of the entire universe are discussed.

But what has this to do with teleportation? The basic operation of teleportation, in the Star Trek sense, can be described as determining the total quantum state of some largish system, transmitting this state information from one place to another, and making a perfect reconstruction of the system at the new location. However, since it is not even in-principle possible to measure the complete state vector of even a very simple quantum system because of the uncertainty principle, as discussed above, this would seem to rule out teleportation on even small quantum systems as physically impossible.

"Not so!" say BBCJPW. There is a way around this quantum roadblock that exploits the peculiarities of EPR nonlocality to transmit the complete description of the state of a quantum system over nature's privileged nonlocal communication channel without performing measurements that extract a complete description of the state vector as information. They propose a multi-step procedure by which any quantum state $| y\rangle$ can be teleported intact from one location to another (but only at a transmission speed that is less than or equal to the velocity of light and only after destroying the initial state). The BBCJPW procedure goes like this:

1. Prepare a pair of quantum sub-systems $| a\rangle$ and $| b\rangle$ in an entangled state, so that they are linked by EPR non-locality.
2. Transport entangled quantum subsystem $| a\rangle$ to the location of the teleport transmitter (the authors call the transmitter operator Alice), and transport subsystem $| b\rangle$ to the location of the teleport receiver (the authors call the receiver operator Bob). These two subsystems are correlated by nonlocality, but at this point contain no information about the quantum state $| y\rangle$ that is to be teleported. In a sense, they represent an open quantum channel that is ready to transmit.
3. Alice brings the to-be-teleported state $| y\rangle$ into contact with the entangled state $| a\rangle$ and performs a set of quantum measurements on the combined system ($| y\rangle$ $| a\rangle$). The details of these measurements have been previously agreed to by Bob and Alice. As a consequence of the measurement process, the original quantum state $| y\rangle$ is destroyed.
4. Using a conventional communication channel, Alice transmits to Bob a complete description of the outcomes of the measurements that she has performed.
5. Bob subjects his quantum subsystem to a set of linear transformations (for example rotations through 90°) that are dictated by the outcomes of Alice's measurements. After these transformations have been done, Bob's quantum subsystem is no longer in state $| b\rangle$ and is now in a state identical to the original quantum state $| y\rangle$, which has in effect been teleported from Alice to Bob.

The BBCJPW scheme for teleportation requires both a normal sub-light-speed communication channel and a nonlocal EPR channel to send the quantum state vector from one location to another, and it also requires considerable pre-arrangement of entangled states and measurement procedures to make the transfer possible. It transfers the quantum system without having completely measured its initial state. The initial state $| y\rangle$ is destroyed at Alice's location and recreated at Bob's location.

BBCJPW analyze the information flow implicit in the process and show that Alice's measurement does not provide any information about the quantum state $| y\rangle$. All of the state information is passed by the "privileged" EPR link between the entangled states. The measurement results can be thought of as providing the code key that permits the EPR-transmitted information to be decoded properly at Bob's end. And because the measurement information must travel on a conventional communications channel, the decoding cannot take place until the code key arrives, insuring that no faster-than-light teleportation is possible.

Quantum teleportation has been experimentally demonstrated in a number of systems over considerable separation distances. Presently, the record distance for successful quantum teleportation is 143 km with photons [2] and 21 m with material systems [3]. Quantum teleportation is also of considerable interest for teleporting information from one quantum memory system to another. In 2012, a Chinese group reported [4] the teleportation of quantum information between two remote atomic-ensemble quantum memory nodes, each composed of 100 million rubidium atoms and connected by a 150-m optical fiber.

The BBCJPW procedure, as mentioned above, is not a design for a machine that teleports macroscopic objects, e.g., human beings, from one location to another. It is concerned with the teleportation only of quantum states of relatively small systems. The complexity of the system increases the number of independent variables that are encoded in the wave function. However, the number of measurements and transformations that must be applied to achieve teleportation only increases as the *natural logarithm* of the number of independent variables describing the system, so a system 1,000 times more complicated would only need 7 times more measurements and transformations.

How complex can a teleported system be? As a guide to answering this question, consider that quantum interference and diffraction effects (see Sect. 6.1) have been observed not only for photons, electrons, helium atoms, and neutrons, but also for increasingly more complex molecules, including the soccer-ball-shaped "buckyball" $C_{60}$ molecule and more recently the molecule "functionalized porphyrin" shown in Fig. 8.1, an extremely complex atomic system with the chemical formula $C_{284}H_{190}F_{320}N_4S_{12}$ [5, 6] and an atomic weight of 10,123 atomic mass units. Apparently quantum interference effects are possible even for very complex systems, as long as the system is cold enough that it does not emit photons while the observation of interference is in progress. This implies (a) that quantum teleportation might also be done with such very complex systems, but (b) that human beings, with bodies that emit large numbers of photons per second in the infrared due to their 98.6° F body temperature, are much too warm to be teleported.

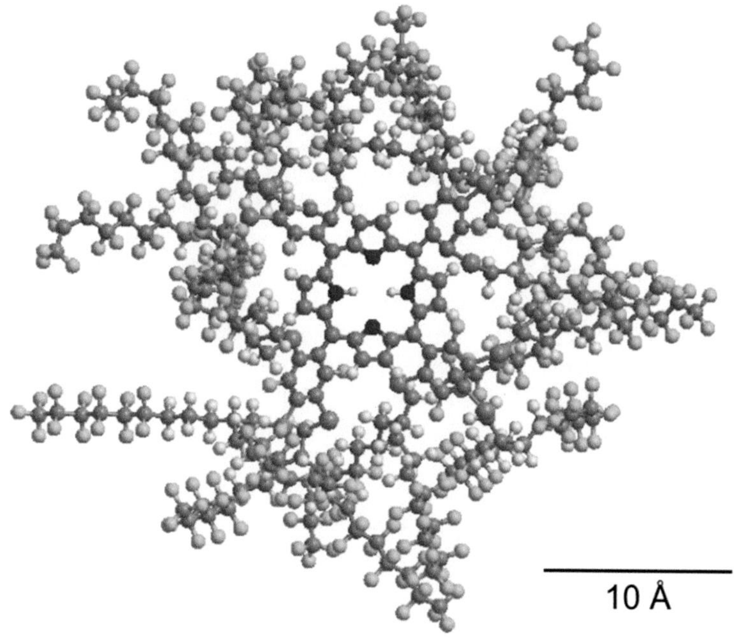


**Fig. 8.1** A molecule of "functionalized porphyrin" $C_{284}H_{190}F_{320}N_4S_{12}$, the most complex molecular system for which quantum interference and diffraction has been demonstrated

The Transactional Interpretation allows us to understand what is going on in the quantum teleportation process. The two entangled sub-systems are in a combined state that is constrained to form an overall completed transaction leading to a final state only when, after appropriate measurements and transformations, the initial state of the "teleportee" system matches the final state of the "teleported" system. At that point the transaction forms and the new system in location B emerges while the old system in location A is destroyed.

## 8.3 Quantum Computing and the TI

In 2014 documents provided by former NSA contractor Edward Snowden showed that the U.S. National Security Agency (NSA) is running an annual $79.7 million research program aimed at developing a quantum computer capable of breaking the encryption of encoded messages and data. So what is a quantum computer?

A quantum computer is a device, first suggested in 1978 by Richard Feynman [7], that uses the manipulation of quantum states to process information and to solve problems, particularly those involving quantum mechanics itself. It is a completely new kind of computer that is presently rising on the technology horizon and that has been reaching significant milestones.

A conventional computer has a memory in which bits of information are stored in definite binary states of 1 or 0, and has a processor that operates on this stored information. For example, the memory may contain two numbers stored in binary representation and a program specifying the locations of these numbers and calling for them to be multiplied together and the result stored. With the execution of the program by the processor, the numbers are fetched, multiplied, and their product in binary representation stored in the memory.

A quantum computer differs from this computing scheme in one important way: the definite-state memory units, the 0-or-1 bits in the conventional computer, are replaced by indefinite *qubits* (short for quantum bits). The qubits are possible states in a quantum system and may be indefinite. For example, all fluorine-19 nuclei have a nuclear spin of $^1/_2$ unit of angular momentum. This spin may point either “up”, i.e., parallel to the flux lines of an external magnetic field, or “down”, i.e. anti-parallel to the direction of the lines of flux. It is a peculiarity of the rules of quantum mechanics that no other spin directions, (e.g., “sideways”) are allowed for $^1/_2$ unit spins, only “up” and “down”. The spin of a fluorine-19 nucleus might be in an up state, a down state, or a mixture of the two, depending on the conditions of preparation. If the spin direction is used as the qubit, as is done in one version of a quantum computer, then the spin orientation represents a binary qubit that may be 1, 0, or a mixture of both.

In this scheme, the up and down spin states have slightly different energies, so a slight addition or subtraction of energy using megahertz radio waves can force the nuclear spins into one orientation or the other. However, it is also possible to prepare the spin orientation of the nucleus in an indefinite state that is 50 % up and 50 % down. There are also techniques for measuring (or reading out) the spin-state qubit as a 0 or 1. If the qubit in an indefinite 50–50 state is read out, the quantum rules require it to randomly “collapse” into either a 1 state or a 0 state, with an equal chance of each outcome.

Of course, a quantum computer needs more than just qubits for its operation. The qubits must be well enough insulated from the random scrambling effects of environmental noise (called decoherence) that the coherent state of the quantum system is preserved for at least long enough to set up a calculation, perform it, and read out the results. It must have the ability to initialize any qubit in a specified state, and to measure the state of a specific qubit. It must have “universal quantum gates”, logical elements capable of arranging any desired logical relationship between the states of qubits. It must also have a processor capable of interlinking such quantum gates to establish rules and boundary conditions for their inter-relationships. In a quantum computation, the arrangement of quantum gates is set by the programmer to connect the qubits in a logical pattern, according to a program or algorithm. After an interval, the qubits assigned to the result are read out. If this is done properly, the quantum computer can be made to perform calculations in a way that is qualitatively different from calculations performed on a conventional computer. In fact, computations with a quantum computer in some cases would require a computation time with a conventional computer greater than the age of the universe.

One powerful capability of a quantum computer is its use in computational problems that require searches over a large number of possibilities in order to find those that satisfy certain criteria. An example is a search over all prime numbers to find a pair that, when multiplied together, produce a specified target number, when that number is very large. This is related to the mathematics of encryption, in which it is easy to multiply two primes together, but extremely difficult to factor the result back into its parent prime numbers. A conventional computer must explore all the combinations of alternatives, one set at a time, in such a search. This may be computationally "hard" if the number of alternatives is very large. A quantum computer with enough qubits to hold the problem, on the other hand, can search the alternatives together in parallel, all at the same time, and reach the problem's solution in a much faster time. In 1994, mathematician Peter Shor devised an algorithm that runs on a quantum computer for integer factorization, and this procedure, called the Shor algorithm, can find the prime factors of large numbers.

In 2001, a prototype 7-qubit quantum computer [8] was constructed using the nuclear spins of seven atoms that are part of a molecule with the iron-based chemical composition $H_5C_{11}O_2F_5Fe$. Two of the eleven carbon atoms in the molecule are carbon isotope 13 and all of the fluorine atoms are fluorine isotope 19. All of the other non-hydrogen atoms in the molecule have even isotope numbers and no nuclear spins. For the quantum computation, a sample of about $10^{18}$ of these molecules is placed in a magnetic field and manipulated using the techniques of nuclear magnetic resonance (NMR), so that the spins function as qubits. NMR is an excellent technology for implementing quantum computing because nuclear spins are well isolated from environmental noise. Therefore, the decoherence time, i.e., the time after which quantum coherence is lost due to random interactions with the environment, is very long. The NMR fields of the quantum computer are manipulated so that Shor's prime factor algorithm is performed and the number 15 is factored into the primes 3 and 5.

To factor a larger number, a system with more than 7 qubits would be required. Things would get interesting when about 24 qubits were available for such computations. With 36 qubits or so, a quantum computer should quickly perform computations that would require a time equal to the age of the universe on a conventional computer. Unfortunately, such a qubit scale-up is difficult, because the NMR device described above is near a hard technology limit of NMR quantum computing. For its operation, all the qubits must be in the same molecule, and molecules with more than 7 or so spins that can be used as qubits do not seem to be feasible.

Fortunately, there are alternative technologies for quantum computing that promise to be more scale-able than the NMR technique. In particular, the technologies of electron spin orientation in quantum dots, nuclear spin orientation of single atom impurities in semiconductors, and manipulation of magnetic flux quanta in superconductors all show promise of providing a basis for scale-able quantum computers. There is intensive work on quantum computing in these areas, but difficult decoherence problems must be overcome before they become competitive with NMR.

What actually goes on in a quantum computer to produce their amazing capabilities? The British mathematician David Deutsch, who has contributed to the development of the theory of quantum computing, likes to invoke the Everett–Wheeler interpretation of quantum mechanics [9, 10] to describe this quantum computing process. Deutsch asserts that the quantum wave function describing the qubits splits among many parallel universes, so that the quantum computer calculations proceed in many parallel universes at the same time. Finally, the answer is found in one of these universes and is transmitted to all the others, which effectively re-collapse into one universe [11]. This is certainly a colorful way of describing quantum computing, but it is very non-economical, in the sense that Deutsch's assertion is in danger of having its throat cut by Occam's Razor.

I prefer to visualize the operation of a quantum computer using the Transactional Interpretation. The programming of the quantum computer that sets up the problem has created a set of conditions such that the quantum mechanical wave function can only form a final transaction and collapse by solving the problem (e.g., finding the prime factors of an input number). The advanced and retarded waves, describing all the possible qubit states of the system, propagate through the computer, seeking the ultimate multi-vertex quantum handshake that solves the problem. This process happens very rapidly, because the propagation time of advanced waves requires a negative time. The net result is that the solution transaction forms, the wave function collapses into the solution state, and the result is read out. This, I think, provides a simple, economical, and insightful view of what goes on in a quantum computer.

How might a quantum computer be put to use? Consider the widely used information protection scheme called RSA public-key encryption, which is embodied in the widely distributed PGP encryption software. Suppose, for example, that Alice wants to use the RSA procedure to send a private message to Bob. Bob has previously posted his public encryption key (a long string of 1s and 0s) in some public database. Alice retrieves Bob's public key from the Internet, and she uses it on her conventional computer running a widely distributed public-key encryption program (e.g. the PGP program) to encrypt her private message to Bob.

Once this is done, the message cannot be decoded, except by Bob, even by someone who has both Bob's public key and Alice's encryption program. Now Bob receives Alice's encrypted message by E-mail and decrypts it using his second private key and the same program. The privacy of the message is ensured because only Bob has the private key, and he has never revealed it to anyone else, not even to Alice.

The RSA encryption algorithm depends critically on the fact that it is easy to multiply two large prime numbers together and obtain their product, but it is computationally "hard" to factor the resulting product back into the original primes, if they are not known. The main part of the public encryption key used by Bob is, in fact, a large number that is the product of two primes, and his private decryption key is one of the prime factors of that public key.

With a conventional computer, it might require many years of computer time to factor the large number in Bob's public key. However, with a quantum computer the large number, at least in principal, can be factored in seconds and the message broken.

Thus, the RSA encryption procedure is based on the assumption of the difficulty of prime factoring, and quantum computers threaten to invalidate that assumption.

It is not surprising that the National Security Agency, the code-breaking and electronic surveillance arm of the U.S. Federal Government, has for many years opposed the wide distribution of strong encryption schemes such as the RSA algorithm. It has also arranged for federal laws that block the distribution in certain forms of strong encryption outside the USA and Canada. It is also not surprising that the NSA is currently one of the principal funding sources for research into quantum computing. Big Brother would like to read our E-mail, even when it is RSA encrypted.

But beyond code breaking, there are other important uses for quantum computers. They should make possible super-fast searches of large databases, for example. Moreover, in physics there are many problems involving the simulation of quantum systems that can only be poorly approximated by calculations on conventional digital computers. With a computer that had a good "quantum co-processor", such physical systems could be simulated much more directly, in ways that are not presently feasible. In my own area of physics research, ultra-relativistic heavy ion physics, we badly need this capability. We limp along without it, making approximations that are only pale shadows of the quantum reality. I would like to have a good quantum computer on my desk, right now.

## References

1. C.H. Bennett, G. Brassard, C. Crépeau, R. Jozsa, A. Peres, W.K. Wootters, Teleporting an unknown quantum state via dual classical and Einstein-Podolsky-Rosen channels. Phys. Rev. Lett. **70**, 1895–1899 (1993)
2. C. Nölleke, A. Neuzner, A. Reiserer, C. Hahn, G. Rempe, S. Ritter, Efficient teleportation between remote single-atom quantum memories. Phys. Rev. Lett. **110**, 140403 (2013), arXiv:1212.3127 [quant-ph]
3. S. Takeda, T. Mizuta, M. Fuwa, P. van Loock, A. Furusawa, Deterministic quantum teleportation of photonic quantum bits by a hybrid technique. Nature **500**, 315–318 (2013)
4. X.-H. Bao, X.-F. Xu, C.-M. Li, Z.-S. Yuan, C.-Y. Lu, J.-W. Pan, Quantum teleportation between remote atomic-ensemble quantum memories, arXiv:1211.2892 [quant-ph]
5. M. Arndt, N. Dörre, S. Eibenberger, P. Haslinger, J. Rodewald, K. Hornberger, S. Nimmrichter, M. Mayor, Matter-wave interferometry with composite quantum objects, in *Atom Interferometry, Proceedings of the International School of Physics "Enrico Fermi"*, vol. 188, ed. by G.N. Tino, M. Kasevich (IOS Press, 2014), arXiv:1501.07770 [quant-ph]
6. S. Eibenberger, S. Gerlich, M. Arndt, M. Mayor, J. Txen, Phys. Chem. Chem. Phys. **15**, 14696 (2013)
7. R.P. Feynman, Simulating physics with computers. Int. J. Theor. Phys. **21**(6), 467–488 (1982)
8. L.M.K. Vandersypen, M. Steffen, G. Breyta, C.S. Yannoni, M.H. Sherwood, I.L. Chuang, Experimental realization of Shor's quantum factoring algorithm using nuclear magnetic resonance. Nature **414**, 6866–8837 (2001)
9. H. Everett, III. Rev. Mod. Phys. **29**, 454 (1957), see also [10]
10. J.A. Wheeler, Rev. Mod. Phys. **29**, 463 (1957), see also [9]
11. D. Deutsch, *The Fabric of Reality: The Science of Parallel Universes—And Its Implications* (Penguin Books, London, 1998). ISBN: 978-0140275414

# Chapter 9
# The Nature and Structure of Time

## 9.1 The Arrows of Time

The Transactional Interpretation describes quantum wave functions traveling in both time directions in an even-handed and symmetric way. From this one is led to ask about the origins of the macroscopic "arrow of time" that is evident in the everyday world. This question is complicated by the presence of at least five seemingly independent "arrows of time" that can be identified in the physical world. Let me briefly review them.

- **The Subjective Arrow of Time**: At the macroscopic level, it is self-evident that the past and the future are not the same. We remember the past but not the future. Our actions and decisions can affect the future but not the past. The past/future distinction is a time arrow.
- **The Electromagnetic Arrow of Time**: We can send electromagnetic signals to the future but not to the past. A current through an antenna makes retarded positive-energy waves but not advanced negative-energy waves.The dominance of retarded electromagnetic waves and potentials is a time arrow.
- **The Thermodynamic Arrow of Time**: Isolated systems have low entropy (i.e., low disorder) in the past and gain entropy and become more disordered in the future. Molecules released from a confining box will rapidly fill a larger volume, but they will not spontaneously collect themselves back in the box. The time direction of entropy increase is a time arrow.
- **The Cosmological Arrow of Time**: The expanding universe was smaller and hotter in the past but will be larger and cooler in the future. The time-direction of expansion is a time arrow.
- **The CP-Violation Arrow of Time**: The $K_L^0$ meson (the neutral long-lived neutral K meson, which is a matter-antimatter combination of a down quark and a strange quark) exhibits weak decay modes having matrix elements and transition probabilities that are larger for the decay process than for the equivalent time-reversed process, in violation of the principle of time-reversal invariance, and represents a



J.G. Cramer, *The Quantum Handshake*, DOI 10.1007/978-3-319-24642-0_9

time arrow. This is related to the so-called CP symmetry violation that is present in the neutral K meson system, which shows a preference for matter over antimatter in certain decay processes. In quantum mechanics, CP is a symmetry transformation involving the simultaneous conversion of matter particles to antimatter and *vice versa* (C) and the reversal of the three spatial coordinate axes (P).

These time arrows cannot be independent, and one would like to understand their connections and their hierarchy. It is generally agreed that the CP violation arrow is probably the most fundamental time arrow, that it was manifested in other particles that were important in the early universe, and that it is responsible for both the dominance of matter in the universe and the breaking of time symmetry in fundamental interactions to make possible the expansion of the universe in a particular time direction, thereby leading to the cosmological arrow of time. From this point on, however, there are at least two divergent views of the hierarchy.

The first of these views might be characterized as the orthodox view of hierarchy, in that it has been widely advocated in the physics literature by many authors, particularly Hawking. [1] It is illustrated by Fig. 9.1a and asserts that the thermodynamic

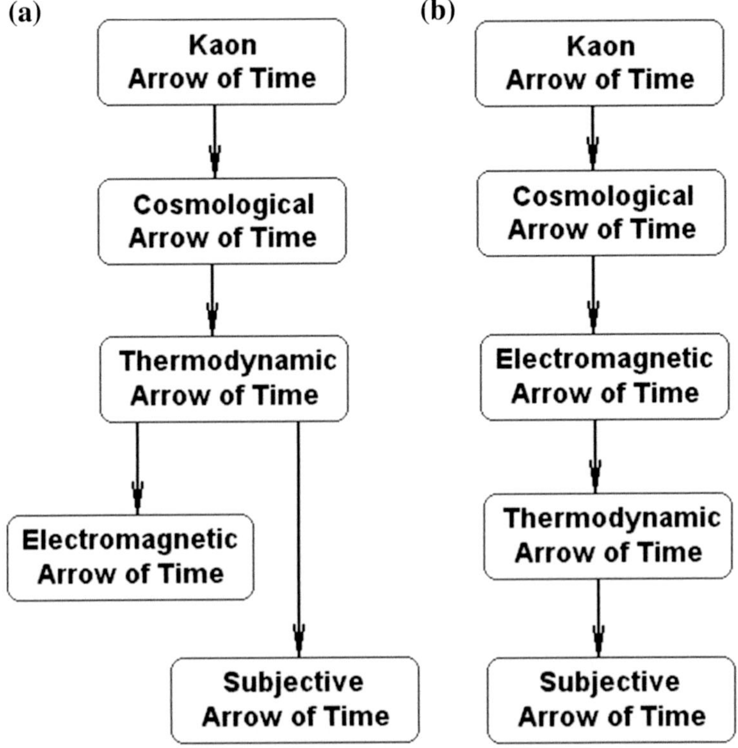


**Fig. 9.1** Two alternative hierarchies of the arrows of time: **a** the hierarchy generally favored in the physics literature; and **b** the hierarchy advocated here

**Fig. 9.2** Ludwig Boltzmann (1844–1906), whose H-Theorem demonstrated the increase in entropy with time

arrow follows from the cosmological arrow and leads to the electromagnetic and subjective arrows of time.

This view goes back to Ludwig Boltzmann (1844–1906) (Fig. 9.2), who in the 1870s "derived" from first principles the thermodynamic arrow of time and the 2nd law of thermodynamics with his famous H-Theorem [2]. If, as Boltzmann appears to demonstrate, the increase in entropy can be derived from first principles using only statistical arguments, this would provide strong support for the primary position of the thermodynamic arrow in the hierarchy of time arrows.

However, the "first principles" that led to Boltzmann's H-Theorem include an implicit time-arrow. Boltzmann used the apparently reasonable assumption that the motions of colliding members of a system of particles are uncorrelated before a collision. Unfortunately, that assumption is not as innocent as it appears. It smuggles into the problem an implicit time asymmetry, which ultimately leads to a system entropy that is constant or increasing with time. However, if one, in the spirit of extending the H-Theorem, had assumed that the motions of colliding particles were uncorrelated *after* the collision, then one would have demonstrated with equal rigor that the entropy was constant or decreasing with time, i.e., the thermodynamic arrow would be pointing in the wrong direction.

This consideration leads to the view of the time arrow hierarchy illustrated by Fig. 9.1b, which asserts that the electromagnetic arrow of time produces the thermodynamic arrow because of the dominance of retarded electrical interactions and electromagnetic waves in the universe. The source of the time asymmetry implicit in the assumptions of the H-Theorem is the intrinsic "retarded" character of electromagnetic interactions between colliding molecules, in which there is a built-in time delay (not a time advance). The term "retarded" means that there is a speed-of-light time delay between the occurrence of some change in a source of electromagnetic field, i.e., the movement of an electric charge, and its appearance in distant electromagnetic fields produced by that source. The field change always occurs *after* the source change, in any reference frame [3]. Thus, the time delays produced by the electromagnetic arrow of time cause the thermodynamic behavior that Boltzmann described.

If a computer simulates with high precision the expansion of gas molecules from a box [4] using instantaneous electromagnetic interactions, suddenly reversing the momentum vectors of the gas molecules sends them back into the box, demonstrating reversibility. If the same simulation is done using retarded electromagnetic interactions, the molecules do not return to the box, and the reversibility is broken. The electromagnetic time-delay builds in an entropy time arrow.

How is the electromagnetic arrow of time related to the cosmological arrow? With their time-symmetric electrodynamics, Wheeler and Feynman [5] sought the answer to this question by arguing for an asymmetry arising from a lower density of absorbers in the past than in the future. However, subsequent authors [6] have argued against this mechanism as inconsistent with an expanding universe. Our own view [7] is that the asymmetry leading to the electromagnetic arrow of time is created by the Big Bang itself, in that advanced waves seeking to go in the negative time direction are stopped, reflected and canceled by the singularity at $T = 0$, which cannot allow advanced waves to penetrate to times before the Big Bang. This is analogous to a shorted transmission, which inverts and reflects electrical pulses seeking to penetrate behind the short. The Big Bang singularity explains the dominance in our universe of retarded electromagnetic radiation and the electromagnetic arrow of time. It also predicts that no Big Crunch singularity lies in our future, for that would similarly suppress retarded waves.

The time arrows listed above did not include a quantum mechanical arrow of time. However, we have observed that from the transactional viewpoint the present source or emitter selects from among the possible quantum transactions based on offer-confirmation echoes, and the future absorber does not. In fact, in a nuclear physics paper [8], I and two colleagues have compared the "post" and "prior" formalisms of the quantum mechanics of nuclear reactions as a way of using the quantum-mechanical arrow of time as an accuracy evaluation criterion in nuclear reaction studies, employing their equivalence as a tool. How does this time preference in the quantum world fit into the above hierarchy? In effect, this quantum mechanical arrow of time is equivalent to the electromagnetic arrow, which requires the dominance of retarded waves. The conditions of our universe favor retarded waves and positive energies and suppress advanced waves and negative energies. While quantum processes may employ the actions of advanced waves to enforce conservation laws, no net "advanced effects" are allowed.

Any hypothetical advanced effects would represent violations of the principle of causality and have not been observed. We conclude that the hierarchy indicated in Fig. 9.1b, the primacy of the electromagnetic/quantum arrow, is consistent with the quantum time preference discussed above and can be used along with a more enlightened version of Boltzmann's H-Theorem to derive the thermodynamic arrow of time.

## 9.2 Determinism and the TI

It has been asserted [9] that the Transactional Interpretation is necessarily deterministic, requiring an Einsteinian block universe to pre-exist, because the future must be fixed in order to exert its influence on the past in a transactional handshake. However, while block-universe determinism is consistent with the Transactional Interpretation, it is not required. A part of the future is emerging into a fixed local existence with each transaction, but the future is not determining the past, and the two are not locked together in a rigid embrace.

Let us make an analogy. The handshakes depicted by the Transactional Interpretation bear some resemblance to the handshakes that take place on Internet lines these days when one uses a debit card to make a purchase at a shop. The shop's computer system reads the magnetic strip on your card and transmits the information to the bank, which verifies that your card is valid and that your bank balance is sufficient for the purchase, and then removes the purchase amount from your bank balance. This transaction enforces "conservation of money". There is a one-to-one correspondence between the amount the store receives for your purchase and the amount that is deducted from your bank account. The transaction assures that the amount is deducted only once, from only one bank account and credited only once, to only one store. On the other hand, the bank does not exert any influence on what you choose to purchase, beyond insuring that money is conserved in the transaction and that you do not overspend your resources.

A quantum event as described by the Transactional Interpretation follows the same kind of protocol. There is a one-to-one correspondence between the energy and other conserved quantities (momentum, angular momentum, spin projections, etc.) that are conveyed from the emitter to the absorber, but aside from the enforcement of conservation laws, the future absorber does not influence the past emission event. "Free will" does not include the freedom to violate physical laws. Therefore, the "determinism" implied by the Transactional Interpretation is very limited in its nature, with plenty of room left over for free choice [10].

## 9.3 The Plane of the Present and the TI

This brings us to the question of how our conceptualization of the plane of the present, our perception of the state of the world at the present instant, is affected by the Transactional Interpretation. In the Transactional Interpretation, the freezing of possibility into reality as the future becomes the present, is not a plane at all, but a fractal-like surface that stitches back and forth between past and present, between present and future.

To make another analogy, the emergence of the unique present from the future of multiple possibilities, in the view of the Transactional Interpretation, is rather like the progressive formation of frost crystals on a cold windowpane. As the frost pattern expands, there is no clear freeze-line, but rather a moving boundary, with fingers of frost reaching out well beyond the general trend, until ultimately the whole window pane is frozen into a fixed pattern. In the same way, the emergence of the present involves a lacework of connections with the future and the past, insuring that the conservation laws are respected and the balances of energy and momentum are preserved.

Is free will possible in such a system? I believe that it is. As previously observed, freedom of choice does not include the freedom to choose to violate physical laws. The transactional handshakes between present and future are acting to enforce physical laws, and they restrict the choices between future possibilities only to that extent.

Therefore, we conclude that the Transactional Interpretation does not require a deterministic block universe. It does, however, imply that the emergence of present reality from future possibility is a far more complex process than we have previously been able to imagine. This is the new transactional paradigm of time.

## References

1. S.W. Hawking, The arrow of time in cosmology. Phys. Rev. D **32**, 2489 (1985)
2. L. Boltzmann, Weitere Studien über das Wärmegleichgewicht unter Gasmolekülen. Sitzungsberichte Akad. der Wiss. **66**, 275–370 (1872)
3. R.E. Kastner, *The Transactional Interpretation of Quantum Mechanics: The Reality of Possibility* (Cambridge University Press, Cambridge, 2012). ISBN: 978-0-521-76415-5
4. J.G. Cramer, Velocity reversal and the arrows of time. Found. Phys. **18**, 1205 (1988)
5. J.A. Wheeler, R.P. Feynman, Rev. Mod. Phys. **17**, 219 (1945)
6. P.C.W. Davies, J. Phys. A **5**, 1025 (1972)
7. J.G. Cramer, Found. Phys. **13**, 887 (1983)
8. R.M. DeVries, G.R. Satchler, J.G. Cramer, Phys. Rev. Lett. **32**, 1377 (1974)
9. J.F. Woodward, Making the universe safe for historians; time travel and the laws of physics. Found. Phys. Lett. **8**, 1–40 (1995)
10. Y. Aharonov, P.G. Bergmann, J.L. Lebowitz, Phys. Rev. **134**, B1410 (1964)

# Chapter 10
# Conclusion

The major failings of many would-be interpretations of quantum mechanics are: (1) they confuse a cause with its effects (*example:* the knowledge interpretation's assumption that changes in knowledge cause the wave function to change); (2) they confuse the map with the territory[1] (*example:* Kastner's assumption in the Possibilist Transactional Interpretation that the wave function resides in the Hilbert-space map of correlations [1]); and (3) they are designed to explain one particular problem while ignoring all others interpretational problems, which we listed in Sect. 2.6 (*example:* the decoherence interpretation's explanation of wave function collapse that ignores the problem of nonlocality). As we have seen, the Transactional Interpretation avoids all of these pitfalls while presenting a comprehensive account of all of the interpretational problems of quantum mechanics.

In Chap. 2 we saw that the peculiar development of quantum mechanics led to an unusual situation in which a well-established formalism that accurately predicted experimental results had no underlying picture that permitted understanding of the mechanisms behind the results. The Copenhagen Interpretation, designed by Bohr and Heisenberg to alleviate this problem, was only partially successful because it relied on positivism and observer knowledge. We saw in Chap. 3 that quantum entanglement and nonlocality, arising from the requirements of conservation laws in the presence of the uncertainty principle, presented severe problems for which the Copenhagen Interpretation offered no answers.

The Transactional Interpretation of quantum mechanics, introduced in Chap. 5, based on advanced/retarded wave handshakes, easily accounts for nonlocality and provides the tools for understanding the many counter-intuitive aspects of the quantum formalism and for visualizing nonlocal quantum processes. The transaction model is "visible" in the quantum formalism itself, once one associates the wave

[1]Consider that (a) I live in Seattle; and (b) Seattle is in the *Rand McNally Road Atlas*; but (c) I do **not** live in the *Rand McNally Road Atlas*.



J.G. Cramer, *The Quantum Handshake*, DOI 10.1007/978-3-319-24642-0_10

function $\psi$ with an offer, the conjugated wave function $\psi^*$ with a confirmation, and quantum matrix elements with completed transactions. In Chap. 5 we described the step-by-step buildup of a transaction, and we showed that the Mead quantum-jump calculation, using the standard quantum formalism, provides a concrete example of how a transaction forms and also provides insights into the origins of randomness in the process.

In Chap. 6 we applied the Transactional Interpretation to a broad range of interpretational problems and paradoxes and saw that in every case the counter-intuitive puzzles presented by the experiments and *gedankenexeriments* were solved by the Transactional Interpretation. In Chap. 7 we investigated the question of nonlocal signals and saw that the quantum formalism contains subtle mechanisms that block potential nonlocal signals and that the Transactional Interpretation is useful in understanding these mechanisms. In Chap. 8 we examined the emerging technologies of quantum communication, quantum teleportation, and quantum computing, and we saw that the Transactional Interpretation is useful in understanding the underlying mechanisms behind these technologies. In Chap. 9 we considered problems related to the nature of time, in particular the origins of the arrows of time and the problem of determinism versus free will, and we saw that the Transactional Interpretation casts new light on these interesting problems.

In summary, we have seen that the Transactional Interpretation of quantum mechanics is unique among the plethora of partial "interpretations" that have arisen in the physics literature over the years in dealing with *all* of the interpretational problems of quantum mechanics and in providing tools for visualizing the mechanisms that lie behind the many puzzling and counter-intuitive results that are emerging from quantum optics laboratories around the world. The Transactional Interpretation gives us a new way of looking at the universe. When you go out at night and look at the stars, consider that not only have the retarded light waves from some star have been traveling for a thousand years to reach your eyes, but also that the advanced light waves from your eyes have reached a thousand years into the past to encourage that star to shine in your direction.

In the Appendices that follow, we will provide further information about the Transactional Interpretation in a question-and-answer format, we will provide a brief introduction to the formalism of quantum wave mechanics, we will introduce games that represents analogies to the quantum correlations that are present in EPR experiments, and we will present more intensively mathematical analyses of several experiments from Chap. 6.

## Reference

1. R.E. Kastner, *The Transactional Interpretation of Quantum Mechanics: The Reality of Possibility* (Cambridge University Press, Cambridge, 2012). ISBN: 978-0-521-76415-5

# Appendix A
# Frequently Asked Questions About Quantum Mechanics and the Transactional Interpretation

## A.1 Basic Questions and Answers

At physics meetings at which talks are given, it is conventional to leave some time at the end of the presentation for questions about ideas that need clarification or that might not have been adequately covered during the formal presentation. Werner Heisenberg remarked [1] that no one who gave a talk at Niels Bohr's Institute in Copenhagen ever finished the presentation, because Bohr always asked too many questions.

So, in the spirit of Bohr's inquisitiveness, here are some questions that may have been raised by material in this book, along with short answers. We will start with a few basic questions.

**Q**: What is quantum mechanics?

**A**: Quantum mechanics is the standard physics theory that deals with the smallest scale of physical objects in the universe, objects (atoms, nuclei, photons, quarks) so small that the lumpiness or *quantization* of otherwise apparently continuous physical variables becomes important.

**Q**: What is the meaning of the angular frequency $\omega$ and wave number $k$ of waves?

**A**: Light waves have a characteristic frequency $f$ indicating how many times per second the electric field of the light wave oscillates. The *angular frequency* $\omega$ is just the same characteristic expressed in radians of phase per second instead of oscillations per second, so $\omega = 2\pi f$. Light waves also have a characteristic wavelength $\lambda$ as they move through space, which is the spatial distance between one electric-field maximum and the next. The *wave number* $k$ is a way of looking at the reciprocal of that characteristic, so that $k = 2\pi/\lambda$. The speed of light $c$ is related to these quantities as $c = f\lambda = \omega/k$. When we deal with particle-waves, for example using the de Broglie wavelength, we also characterize them in terms of $\omega$ and $k$.

**Q**: What is quantization?



J.G. Cramer, *The Quantum Handshake*, DOI 10.1007/978-3-319-24642-0

**A**: It's the idea that there are minimum size chunks for certain quantities like energy and angular momentum. The minimum energy chunk for light of frequency $f$ is $E = hf = \hbar\omega$, where $h$ is Planck's constant and $\hbar$ is $h$ divided by $2\pi$. We call the particle of light carrying this minimum-size energy chunk $\hbar\omega$ a photon.

**Q**: How big is Planck's constant?

**A**: Planck's constant is very small. It represents the minimum unit of action (energy $\times$ time) in quantum physics, and it has the value $h = 6.62606957 \times 10^{-34}\ J \cdot s = 4.135667516 \times 10^{-15}\ eV \cdot s$. Further, $\hbar = h/2\pi = 1.054571726 \times 10^{-34}\ J \cdot s = 6.58211928 \times 10^{-16}\ eV \cdot s$. As an example of the use of $h$, red visible light has a wavelength of about $\lambda_\gamma = 500$ nanometers (nm). A photon of red light carries an energy of $E_\gamma = hc/\lambda_\gamma = 2.48\,eV$. An $eV$ is an electron-volt, the energy required to move one charged electron through a potential of 1 V. An electron-volt is a very small quantity of energy, roughly the amount of energy that an atomic electron gains or loses by jumping from one atomic orbit to another.

**Q**: What if Planck's constant was zero?

**A**: Setting $\hbar$ to zero is a way of taking the classical limit of quantum mechanics: the uncertainty principle goes away, and in that limit quantum mechanics becomes more or less equivalent to Newtonian mechanics. However, Planck's treatment of black body radiation suggests that a universe in which Planck's constant was zero would be very different than ours, because hot objects would rapidly radiate away all their energy as high-frequency electromagnetic radiation (the ultraviolet catastrophe), matter would be extremely cold, and the universe would be dominated by light.

**Q**: What is angular momentum, why is it quantized, and why is it conserved?

**A**: Angular momentum is the rotational momentum of a rotating or spinning object. It is the product of the angular rotation rate $\omega$ and the rotational inertia $I$ of the object. In quantum mechanics, if one requires the state of an object that is slowly turned through 360° to return to exactly the same state as before, this has the rather unexpected consequence of requiring that angular momentum must be quantized in units of $\hbar$. Thus, it is 360° rotational symmetry that causes angular momentum to be quantized. Therefore, one would think that no object in our universe could have a non-zero angular momentum of less than one $\hbar$ unit. As it turns out, this is wrong, because fermions (see below) have an intrinsic spin of *half* an $\hbar$ unit and therefore do not have 360° symmetry. Angular momentum conservation is the rotational manifestation of Newton's 3rd Law, which in it's rotational form states that in the absence of external torques, the angular momentum of a system is an unchanging time-independent constant.

**Q**: What is the difference between orbital angular momentum and spin angular momentum?

**A**: In an atom, two distinct kinds of angular momentum are present: the angular momentum created by the electrons orbiting the nucleus and the intrinsic spin angular momentum of the electrons themselves. The orbital angular momentum always comes in units of $\hbar$, while the electron spins are each $\hbar/2$. The total angular momentum is the vector sum of these components, and it can be quite complicated because of the variety of ways in which the component angular momentum vectors can couple. The parity (see below) of the system wave function depends on the orbital angular momentum, with odd values $(1, 3, 5, \ldots)$ giving odd parity and even values $(0, 2, 4, \ldots)$ giving even parity.

**Q**: What is parity?

**A**: A parity transformation means that the three spatial coordinates ($x$, $y$, and $z$) of a quantum system are reversed in sign and direction. If the quantum wave function remains the same under such a transformation, the system is said to have positive parity. Example: $\cos(x) = \cos(-x)$. If the quantum wave function changes sign under such a transformation, the system is said to have negative parity. Example: $\sin(x) = -\sin(-x)$. Each energy state of an atom or nucleus has a definite parity, either positive or negative. In a "quantum jump" atomic transition, if the parity changes the properties of the emitted photon will be different than if the parity remains the same. Parity is related to mirror-image symmetry.

**Q**: Why is photon linear polarization, important in EPR experiments, constrained by angular momentum conservation?

**A**: Photons are boson particles (see below) that have an intrinsic spin angular momentum of one $\hbar$ unit. If this spin vector points in the direction of motion of the photon, the photon is in a state of left circular polarization; if the spin vector points against the direction of motion, it is in a state of right circular polarization. Because the photon travels at the speed of light, special relativity does not allow its spin to point in any other directions. States of linear polarization can be formed by a superposition of the right and left circular polarization states, as discussed in Sect. 6.6 and quantified by Eqs. 6.4 and 6.5. Since angular momentum conservation constrains the spins of the photons of a system, it also constrains the states of linear polarization, and EPR experiments make use of this.

**Q**: What is a fermion?

**A**: Fundamental particles called *fermions*, including electrons, muons, neutrinos, and quarks, have an intrinsic angular momentum or "spin" of ½ $\hbar$, i.e. half an $\hbar$ unit. Composite fermions, which are made up of three quarks, include neutrons, protons, and heavier baryons. Fermions obey Fermi-Dirac statistics and the Pauli exclusion principle, the requirement that only one fermion can occupy any particular quantum state. This leads to the electron-shell structure in atoms, to the neutron and proton shell structures in nuclei, and to valence behavior and chemical bonding in chemistry. In quantum wave mechanics, fermions are described using the Dirac wave equation.

The half-integer spin of fermions means that one must rotate such particles through two full revolutions or 720° before they return to their original state. These peculiar fermion particles also come with antimatter twins (e.g., positrons for electrons and anti-quarks for quarks) of opposite parity (opposite mirror-symmetry).[1]

**Q**: What is a boson?

**A**: Fundamental particles called *bosons*, including photons, gluons, the $Z^0$ and $W^{\pm}$ weak-interaction particles, and the Higgs particle, the mediating particles for the fundamental forces, all have an intrinsic spin that is either zero or an integer multiple of $\hbar$. Composite bosons include alpha-particles, many atoms and nuclei, and $\pi$- and $K$-mesons. Bosons obey Bose–Einstein statistics and can form a "condensate" in which a large number of bosons all have identical wave functions and occupy the same quantum state. Lasers are an example of the operation of Bose–Einstein statistics for photons. In quantum mechanics, bosons are described by the electromagnetic wave equation for light and the Klein–Gordon equation for massive particles.

**Q**: Is light made of particles or of waves?

**A**: Light exhibits the behavior of both a particle and a wave. The Transactional Interpretation shows us that light moves from place to place as waves, but at locations where it is emitted or absorbed it obeys the boundary conditions of a particle carrying energy $\hbar\omega$, momentum $\hbar k$, and angular momentum $\hbar$. One might say that light travels as a wave but takes-off and lands as a particle. It is usually convenient to think of low energy photons, e.g., radio transmissions, as waves, and high energy photons like X-rays and gamma rays as particles. In the TI the wavelike-behavior is present in the offer and confirmation waves, and the particle-like behavior is present in the completed transactions.

**Q**: Are electrons really particles?

**A**: Electrons follow much the same rules as photons, traveling as waves but taking-off and landing as particles. The Hanbury-Brown-Twiss effect discussed in Sect. 6.9 demonstrates that particles in transit do not have a separate identity from their wave functions, and that there can be no one-to-one correspondence established between a particle that is emitted and a particle that is subsequently captured or absorbed. One can "manufacture" an arriving particle from fractions of the particles that departed. Further, the Bohr/de Broglie view of electrons orbiting in atoms is that they are standing waves bent around the circumference of the orbit. Nevertheless, it is convenient to think of electrons as particles, because the particle aspects of their nature are normally much more apparent that their wave aspects.

**Q**: What is the difference between interference and diffraction?

**A**: Both are wave phenomena. Diffraction is the behavior of a wave that has had part of its extent "chopped off" by a slit or aperture. When a wave consisting of wave fronts that are parallel planes (i.e., a plane wave) comes to a pinhole, only that part of the wave that overlaps the hole can get through. The result is that the emerging wave fronts are curved surfaces like hemispheres centered on the pinhole.

---

[1] I have always suspected that there must be a connections between the peculiar fractional spin of fermions and their antimatter and parity behavior, but no one has ever been able to explain it to me, including prominent string theorists who claim to be masters of a "theory of everything".

Attempting to restrict the wave to a small region causes it to "diffract" and spread out as it propagates further.

Interference comes from the combination of two or more waves of the same wavelength and frequency that overlap with some definite phase relation between them. If two waves have a phase difference of 0 or some integer multiple of $2\pi$, the interference will be *constructive*, and the resulting wave will have twice the amplitude of its components. If two waves have a phase difference that is some odd-integer multiple of $\pi$, the interference will be *destructive*, and the resulting wave will cancel and have zero amplitude. A good example of this kind of behavior is two-slit interference as discussed in Sect. 6.1.

**Q**: What is the difference between an experiment and a *gedankenexperiment*?

**A**: Many in-principle experimental ideas, i.e., demonstrations of natural processes in an experimental setting where measurements are made, are either too difficult to implement in practice or are so obvious that they are not worth actual implementation, and therefore fall into the category of thought experiments or *gedankenexperiments*. Examples of the too-difficult variety are the boxed-atom experiments of Sect. 6.19. Examples of the too-obvious variety are Einstein's Bubble (Sect. 6.2), Schrödinger's Cat (Sect. 6.3), and Renninger's Negative-Result Experiment (Sect. 6.6). In the discussion of applications of the TI to experiments, we have labeled the real experiments that have actually been performed in the quantum optics laboratory with an asterisk (*) to distinguish them from the *gedankenexperiments*.

**Q**: What is meant by "the formalism" of quantum mechanics?"

**A**: Basically, the formalism of quantum wave mechanics is mathematics consisting of (1) a differential equation like Schrödinger's wave equation that relates mass, energy, and momentum; (2) the mathematical solutions of that wave equation, called wave functions, which contain information about location, energy, momentum, etc. of some system; (3) operators that can extract quantities of interest from quantum wave functions, and (4) procedures for using operators and wave functions to make predictions about physical measurements on the system. See Appendix B for more details. The formalism of quantum matrix mechanics involves using matrix representations and manipulations to describe and connect the states of a system and will not be described further here.

**Q**: What is a "system" in quantum mechanics?

**A**: A quantum system is any collection of physical objects that is to be described by a wave function. It could be a single electron, a group of quarks, an atom, a cat in a box, a quantum computer, or the whole universe and all its contents.

**Q**: What is meant by "entanglement"?

**A**: Entanglement is a term coined by Schrödinger to indicate that the quantum state of one particle depends on some details of the quantum state of the other particle. Entanglement often occurs because two entangled particles are emitted by the same source, and some conservation law, e.g., energy, momentum, or angular momentum conservation, can only be preserved if the particles have values of that quantity that are correlated. See Sect. 3.3 for a discussion of entanglement.

**Q**: What is quantum nonlocality?

**A**: Albert Einstein distrusted quantum mechanics because he perceived embedded in its formalism what he called "spooky actions at a distance". The characteristic that worried Einstein is now called "nonlocality". The term locality means that separated system parts that are out of speed-of-light contact can only retain some definite relationship through memory of previous contact. Nonlocality means that some relationship is being enforced nonlocally across space and time. The nonlocality of quantum mechanics has been spotlighted by quantum-optics EPR (Einstein–Podolsky–Rosen) experiments showing in ever increasing detail the peculiar actions and consequences of nonlocality. These measurements, for example the correlated optical polarizations for oppositely directed photons, show that something very like faster-than-light hand-shaking must be going on within the formalism of quantum mechanics and in nature itself.

**Q**: Can quantum nonlocality be used to send nonlocal signals faster than light and backwards in time?

**A**: This question is addressed in some detail in Chap. 7. The short answer is that nonlocal signaling is impossible unless the standard formalism of quantum mechanics is an approximation to a more general theory that might allow such signaling. Several philosophers of science [2–7] have demonstrated that the formulators of the standard quantum formalism used the impossibility of nonlocal signals as a guide in constructing the theory and thereby built that impossibility into the formalism. It is not clear if there might be a more general quantum formalism in which this built-in prohibition is eliminated. Thus, the doorway to nonlocal signaling is open just a crack, but it is a very small crack.

**Q**: What is an interpretation of quantum mechanics?

**A**: Quantum mechanics arrived in the world of physics without an interpretation that allowed one to picture what was going on behind the mathematics. See Chap. 2. Since that time, starting with the Copenhagen Interpretation devised in the late 1920s, there have been a number of attempts to solve this problem. The subject of this book, the Transactional Interpretation, is one of these. Many ideas that are promoted as "interpretations" have the deficiency that they only focus on one or two of the interpretational problems of the quantum formalism listed in Sect. 2.6. The Transactional Interpretation deals with all of them.

**Q**: Why not use experimental results to determine the correct interpretation of quantum mechanics?

**A**: The problem is that, unlike most ideas in physics, QM interpretations cannot be tested, verified, or falsified by laboratory experiments because all of the rival interpretations are describing the same mathematics, and it is the mathematics that makes the testable predictions. Interpretations of QM can only be eliminated if (a) they fail to explain quantum phenomena,[2] or (b) they are found to be inconsistent with the QM mathematics. See Sects. 6.15 and 6.20 for examples of the latter situation.

**Q**: Why is there so much controversy about interpreting quantum mechanics?

**A**: Interpretations of quantum mechanics cannot be falsified by experimental tests, and so there is no good way of eliminating the inadequate ones. The problem of QM interpretation is compounded because there are no social forces propelling the physics and philosophy communities to settle on one interpretation and adopt it as the standard, as is normally done with testable physical theories. Rather the social forces work in the other direction, giving rewards in the form of recognition, conference invitations, and tenure to those who "do their own thing" in the area of QM interpretations, since it is a playground where their ideas cannot be tested or falsified. There seems to be more prestige in having your own interpretation than in adopting someone else's. In lieu of testing interpretations by performing experiments, the philosophy-of-science community seems to have devolved to "challenges" in which advocates of each interpretation attempt to poke holes in the interpretations of their rivals. This is rather like horseback-mounted knights engaging in jousting duels in a medieval court. It may be entertaining to spectators and some participants, but it does not promote convergence.[3]

**Q**: What is the Copenhagen Interpretation of quantum mechanics?

**A**: The Copenhagen Interpretation of quantum mechanics is a set of ideas and principles devised by Bohr, Heisenberg, and Born in the late 1920s to give meaning to the formalism of quantum mechanics and to avoid certain "paradoxes" that seemed implicit in the formalism. It is the interpretation that may be presented in textbooks on quantum mechanics. It is described in detail in Sect. 2.6 above. Its principal elements are: Heisenberg's uncertainty principle, Born's probability rule, Heisenberg's knowledge interpretation, and Bohr's complementarity principle.

We should note that the Copenhagen Interpretation, first explicitly given that name by Heisenberg in 1955 [8], is somewhat of a moving target, since both Bohr and Heisenberg modified their views during their long careers and they often disagreed over the finer points of the interpretation. In particular, according to von Weizsäcker [9] Heisenberg considerably modified his emphasis on logical positivism over the years.

---

[2]In fact, many "interpretations" that fail to explain quantum nonlocality, particularly those that, like the decoherence interpretation, focus exclusively on wave function collapse, remain alive and well and continue to attract adherents.

[3]As a personal note, my own work on the Transactional Interpretation was done well after I had been granted tenure at the University of Washington and probably has had a net negative impact on my academic career, since my DOE-funded primary research area, experimental nuclear and relativistic heavy-ion physics, is very far removed from quantum interpretations and the philosophy of science.

**Q**: What is Heisenberg's uncertainty principle?

**A**: In the quantum formalism there are pairs of variables, including conserved quantities, that are multiplied together in the wave functions and that Bohr has called "complementary". These pairs include position/momentum, energy/time, and angular position/angular momentum. These variables are subject to Heisenberg's uncertainty principle, which describes a "see-saw" relation between the uncertainties in the two members of the pair. For example, the uncertainty in the position on the $x$-axis $\Delta x$ and the uncertainty in the $x$-component of momentum $\Delta p_x$ are governed by the uncertainty relation $\Delta x \Delta p_x \geq \hbar$. If $\Delta x$ is decreased, then $\Delta p_x$ necessarily increases, and *vice versa*. As discussed in Sect. 2.4, this behavior can be explained in terms of the mathematics of Fourier analysis. It is important to realize that the mathematics of Fourier analysis tells us that if a particle has a well-defined value for one of two complementary variables, it cannot have a well-defined value for the other variable, hidden or not.

**Q**: What is Born's probability rule?

**A**: The Born probability rule is the method for extracting a probability from quantum wave functions. The probability is $P = \psi\psi^*$ where $\psi$ is the wave function describing the state of a quantum system and $\psi^*$ is the complex conjugate or time-reverse of the wave function. This forces the probability to be real and positive, even when the wave function is a complex quantity, with real and imaginary parts. For the Copenhagen Interpretation, the Born rule is an assumption, a non-obvious postulate that must be adopted *ad hoc*. One of the major arguments in favor of the Transactional Interpretation is that the Born rule is a direct consequence of the transaction model and not an independent assumption.

**Q**: What is Heisenberg's knowledge interpretation?

**A**: Heisenberg's knowledge interpretation is the assertion that the wave function of quantum mechanics is a mathematical encoding of the knowledge of an observer who is making measurements on the system that the wave function describes. This assumption offers an explanation of how the wave function collapses (the observer gains knowledge from a measurement) and deals with simple nonlocality problems like the Einstein's bubble paradox described in Sect. 6.2. It may seems strange that the solution to a simple second-order differential equation relating mass, energy, and momentum is supposed to provide a map of thought processes of an observer, but that is what the knowledge interpretation asks us to accept, and indeed it has been accepted by most of the prominent theorists of the late 20th century. From the point of view of the Transactional Interpretation, the knowledge interpretation has it backwards: the change in observer knowledge is a consequence of a change in the wave function (because a transaction forms), not the cause of the wave function change.

**Q**: What is Bohr's complementarity principle?

**A**: Complementarity is a term that Nies Bohr invented to describe and generalize the behavior of complementry variables in a quantum system. Bohr [10] described complementarity in these words:

> …however far the [quantum] phenomena transcend the scope of classical physical explanation, the account of all evidence must be expressed in classical terms. The argument is simply that by the word "experiment" we refer to a situation where we can tell others what we have done and what we have learned, and that, therefore, the account of the experimental arrangements and of the results of the observations must be expressed in unambiguous language with suitable application of the terminology of classical physics.
>
> This crucial point… implies the impossibility of any sharp separation between the behavior of atomic objects and the interaction with the measuring instruments which serve to define the conditions under which the phenomena appear.… Consequently, evidence obtained under different experimental conditions cannot be comprehended within a single picture, but must be regarded as complementary in the sense that only the totality of the phenomena exhausts the possible information about the objects.

Essentially, complementarity is the idea that two seemingly contradictory descriptions can characterize the same phenomenon. It is a generalization from the uncertainty principle to the ideas that there are balanced "unknowables" in quantum systems and that a system and measurement on that system are inseparable parts of the whole and cannot be considered separately. In his writings over the years, Bohr moved from using the wave versus particle dichotomy as an example of complementarity to using the more general example of comparing dynamic principles to teleological conservation-law principles [11].[4]

**Q**: What is the Many-Worlds Interpretation of quantum mechanics?
**A**: Hugh Everett III (1930–1982), a student of John Wheeler at Princeton, introduced the many-worlds interpretation (MWI) in an attempt to explain the curious QM phenomenon of wave function collapse [12, 13]. Everett asserted that in a measurement of some physical quantity with several possible outcomes, the wave function does not collapse at all, but instead the universe itself splits into alternative universes, each of which contains one of the possible outcomes of the measurement. Quantum interference between universes can occur when they are indistinguishable. This peculiar view of quantum phenomena has accumulated many adherents. Notable among these is David Deutsch [14], who views the MWI as an aid to thinking about quantum computers as parallel processing in many parallel universes at once. Problems with the MWI are: (1) that it is very confused about the origin and operation of the Born probability rule, and (2) that it does not seem to be able to explain quantum nonlocality. In fact, Everett described the EPR work as a "false paradox" and promised to write a subsequent paper that would explain how to deal with it using the MWI [12]. He never made good on that promise.

---

[4]I was on the second row of a large lecture hall when Niels Bohr, near the end of his life, gave a Physics Colloquium at Rice University in 1960. He recounted for us his solution of the Einstein Clock paradox at the 8th Solvay Conference in 1930, and he spoke at length on the power of the principle of complementarity. His Danish-accented English was somewhat difficult to understand at first, but it was a very interesting talk.

**Q**: Can the quantum wave function $\psi$, particularly for multi-particle systems, be considered a real object that exists in normal three-dimensional space?

**A**: Yes, in the TI view the quantum wave function $\psi$ can be taken as a real-but-incomplete object moving through ordinary 3-dimensional space. See Sect. 5.7. The wave function is the offer wave that initiates a quantum transaction that transfers energy and momentum across space-time. In the multi-particle case, multi-vertex transactions form that select from the "free" (i.e., uncorrelated) wave functions of each particle only those wave function components that satisfy appropriate conservation laws at all of the vertices, thereby enforcing entanglement.

**Q**: In Chap. 7 on nonlocal signaling, it was demonstrated that the potential signal is blocked by the superposition of a switchable interference pattern and an "anti-interference" pattern, the latter 180° out of phase with the former, so that it "erases" the signal. Why not just shift the phase of the anti-interference pattern by 180°, so that it does not do this?

**A**: There is no general answer to this question, but for the experimental setups discussed in Chap. 7 there is no place that a phase shift element, e.g., a half-wave plate, could be inserted so that it would shift the phase of one of the interference patterns without shifting the other by an equal amount. There does not seem to be a way of unblocking the blocker.

**Q**: Why is it the emitter that is responsible for transaction formation?

**A**: The emitter chooses, weighted by echo strength, which of the offer/confirmation echoes that it receives from potential absorbers is to be used as the trigger to form a transaction with that absorber, and it may also choose to form no transaction at all.

**Q**: That is not time-symmetric; why is it the emitter rather than the absorber that selects the transaction"

**A**: There is a quantum mechanical arrow of time that favors the emitter over the absorber. It is related to the electromagnetic arrow of time that favors retarded over advanced electromagnetic waves. This is discussed further in Sect. 9.1.

**Q**: What is "pseudo-time" in the Transactional Interpretation?

**A**: The Transactional Interpretation account of the formation of a transaction is layered, with an offer wave followed by a confirmation wave followed by echo selection followed by transaction formation. However, since these processes are occurring in both time directions, the layers cannot represent a time sequence in the usual sense. Therefore, we say that the sequence occurs in pseudo-time, meaning the layered causal sequence in the build-up to a transaction.

**Q**: How can the Transactional Interpretation account for the interaction of two wave functions, for example, those involving only spins?

**A**: This question is one for the quantum formalism, not the interpretation of it. The wave functions that the TI describes are generated from a wave equation containing a Hamiltonian that characterizes the interactions present in the system that is being described mathematically. Thus, the wave functions, as offer and confirmation waves, have the interactions already built into them by the formalism that they describe.

**Q**: Why doesn't the Transactional Interpretation provide a detailed mathematical description of the formation of a transaction?

**A**: A proper interpretation describes the formalism and does not modify or elaborate it. Otherwise, it is a new theory, not an interpretation. Therefore, it would be inappropriate for the Transactional Interpretation to provide a detailed mathematical description of the formation of a transaction. However, we note that Carver Mead [15] has provided a quantum mechanical description of the "quantum jump" transport of a photon from one atom to another using mixed atomic states, and this can be taken as a prototype for the formation of a transaction as an exponential avalanche triggered by perturbations from an exchange of advanced and retarded waves. His calculation is described in Sect. 5.4 above.

**Q**: How can the Transactional Interpretation be applied to relativistic quantum mechanics?

**A**: There are two branches of relativistic quantum mechanics, relativistic wave mechanics and quantum field theory. The Transactional Interpretation is fully compatible with relativistic wave mechanics and indeed depends on the advanced wave solutions of relativistic wave equations.

Ruth Kastner, in her book on the TI [16] goes into some detail on the application of the Transactional Interpretation to quantum field theory. She provides some new and interesting insights into quantum processes at the Feynman-diagram level and encounters no problems in doing so. However, in our opinion quantum field theory is suspect because it predicts that the energy content on the quantum vacuum is $10^{120}$ times larger than its actual value. We feel that quantum field theory must eventually be replaced by a new theory of quantum gravity that will accommodate both particle physics and gravitational physics. Therefore, placing emphasis on developing any detailed interpretation of the present state of quantum field theory is problematical, since it must inevitably change.

**Q**: Why should one accept an interpretation that has waves going backwards in time, when nothing in the real world behaves that way?

**A**: Two reasons: (1) it seems to be the only way of providing any visualizable explanation of quantum entanglement and nonlocality, and (2) the advanced waves that go backwards in time, as complex-conjugated wave functions $\psi^*$, are present in the standard formalism of quantum wave mechanics for all to see, if they just open their eyes and their mind.

## A.2 Questions from Chapter 3 on Nonlocality

These are questions taken from Chap. 3, which discusses quantum nonlocality:

**Q**: Can the quantum wave functions of entangled systems be objects that exist in normal three-dimensional space?
**A**: Yes. The wave functions are objects in normal three-dimensional space, and the entanglement and nonlocal correlations between objects are arranged by the application of conservation laws at the vertices of the final transaction.

**Q**: What are the true roles of the observers and measurements in quantum processes that involve several separated measurements on entangled subsystems?
**A**: The observer and measurement can be end-points of a transaction, but they have no special status. Giving "observer knowledge" a special role amounts to confusing the effect with the cause in wave function collapse.

**Q**: What is wave function collapse (or state vector reduction) and how does it occur, particularly for entangled systems?
**A**: Wave function collapse and state vector reduction can be understood as the formation of a transaction that projects out one possible outcome from among the possibilities implicitly or explicitly contained in the initial offer wave function. For entangled systems a multi-vertex transaction forms in which conservation laws are enforced at all vertices together to enable entanglement and nonlocal correlations.

**Q**: How can quantum nonlocality be understood?
**A**: Nonlocality can be understood as arising from a muli-vertex transactional handshake of retarded and advanced waves that connects and correlates entangled objects and their source across space-time.

**Q**: How can quantum nonlocality be visualized?
**A**: Nonlocality can be visualized as the multi-vertex transaction across space-time, connecting and correlating separated system parts. Chapter 6 provides many examples of such multi-vertex transactions.

**Q**: What are the underlying physical processes that make quantum nonlocality possible?
**A**: The underlying physical process that makes quantum nonlocality possible is the transactional handshake between retarded and advanced waves across space-time that connects entangled and separated parts of an overall system back to the event at which they separated, so that conservation laws appropriate to the system can be enforced.

## A.3 Questions from Discussions of the TI Found on the Internet

These are questions taken from various "challenges to the TI" extracted from Internet-based discussions of the Transactional Interpretation.

**Q**: "Isn't the Transactional Interpretation mathematically imprecise?"

**A**: This is a peculiar question, because the mathematics of quantum mechanics is contained in the formalism that the interpretation describes, not in the interpretation itself. That said, the Transactional Interpretation does not have any problems with mathematical precision. The offer waves $\psi$ for massive particles are solutions of the Schrödinger wave equation and the confirmation waves $\psi^*$ are solutions of the complex conjugate of the Schrödinger wave equation, both derivable from relativistically-invariant wave equations. A transaction formation (see Sect. 5.6) is a genuinely stochastic event, and therefore does not obey a deterministic equation. Outcomes based on actualized transactions obey the Born rule, and the TI provides a derivation of the Born probability rule rather than assuming it, as is the case in the Copenhagen Interpretation. The TI is a direct map of the quantum formalism and is imprecise only to the extent that the underlying formalism is imprecise.

**Q**: "Can the Transactional Interpretation make new predictions that are experimentally testable, so that it can be verified or falsified?"

**A**: No consistent interpretation of quantum mechanics can be tested experimentally, because each is an interpretation of the same quantum mechanical formalism, and the formalism makes the predictions. The Transactional Interpretation is an exact interpretation of the QM formalism. Like the Many-Worlds and the Copenhagen interpretations, the TI is a "pure" interpretation that does not add anything *ad hoc*, but does provide a physical referent for a part of the formalism that has lacked one (e.g., the advanced wave functions appearing in the Born probability rule and amplitude calculations). Thus the demand for new predictions or testability from an interpretation is based on a conceptual error by the questioner that misconstrues an interpretation as a modification of quantum theory. According to Occam's Razor, the hypothesis that introduces the fewest independent assumptions is to be preferred. The TI offers this advantage over its rivals, in that the Born probability rule is a result rather than an independent assumption.

**Q**: "Just where and when in space-time does a transaction occur?"

**A**: A clear account of transaction formation is given in Chap. 5, which pictures a transaction as emerging from an offer-confirmation handshake as a four-vector standing wave in normal three-dimensional space with endpoints at the emission and absorption vertices. Kastner [16] has presented an alternative account of transaction formation in which the formation of a transaction is not a spatiotemporal process but one taking place on a level of possibility in a higher Hilbert space rather than in normal three-dimensional space.

**Q**: "Hasn't Maudlin [17] already demonstrated that the Transactional Interpretation is inconsistent?"

**A**: No, he has not. Maudlin's *gedankenexperiment* was discussed in Sect. 6.14. Maudlin [17] raised an interesting challenge for the Transactional Interpretation by pointing out a paradox that can be constructed when the non-detection of a slow particle moving in one direction modifies the detection configuration in another direction. This problem is dealt with by the TI in Sect. 5.5 above by introducing a hierarchy in the order of transaction formation. Maudlin's paradox is discussed in detail in Sect. 6.14. Other solutions to the problem raised by Maudlin can be found in references [18–20].

**Q**: "Hasn't Maudlin [17] demonstrated that the Transactional Interpretation leads to an incorrect calculation of probability?"

**A**: Maudlin did make that incorrect claim. It is based on his assumption that the wave function is a representation of observer knowledge and must change when new information becomes available. That Heisenberg-inspired view is not a part of the Transactional Interpretation, and introducing it leads to bogus probability arguments. In the Transactional Interpretation, the offer wave does not magically change in mid-flight at the instant when new information becomes available, and its correct application leads to the correct calculation of probabilities that are consistent with observation.

**Q**: "Hasn't Maudlin [17] demonstrated that the Transactional Interpretation is deterministic rather than stocastic?"

**A**: Maudlin certainly made that claim, but it has no basis. The intrinsic randomness of the Transactional Interpretation comes in the third stage of transaction formation, in which the emitter, presented with a sequence of retarded/advanced echoes that might form a transaction, hierarchically and randomly selects one (or none) of these as the initial stage of transaction formation, as described in Sect. 5.5.

In Carver Mead's TI-based mathematical description of a quantum jump [15], the perturbations between emitter and absorber create a frequency-matched pair of unstable mixed states that either exponentially avalanche to a full-blown transaction with the transfer of energy or else disappear due to boundary conditions when a competing transaction forms. In a universe full of particles, this process does not occur in isolation, and both emitter and absorber also experience random perturbations from other systems that can randomly drive the instability in either direction. Ruth Kastner [16] has described this situation in transaction formation as "spontaneous symmetry breaking", which relates the process to analogous behavior in quantum field theory.

**Q**: "How can the Transactional Interpretation handle the quantum mechanics of systems with more than one particle?"

**A**: This is a peculiar question, because Refs. [21, 22] long ago provided many examples of the application of the TI to systems involving more than one particle.

These include the Freedman-Clauser experiment, which describes a 2-photon transaction with three vertices (see Sect. 6.8), and the Hanbury-Brown-Twiss effect, which describes a 2-photon transaction with four vertices (see Sect. 6.9). Chapter 6 above provides many examples of more complicated multi-particle systems, including systems with both atoms and photons.

But perhaps the question posed above is based on the belief that quantum mechanical wave functions for systems of more than one particle cannot exist in normal three-dimensional space and must be characterized instead as existing only in an abstract Hilbert space of many dimensions. Indeed, Kastner's "Possibilist Transactional Interpretation" [16] takes this point of view and describes transaction formation as ultimately appearing in 3D space but forming from the Hilbert-space wave functions.

As discussed in Sect. 5.7, the "standard" Transactional Interpretation presented here, with its insights into the mechanism behind wave function collapse through transaction formation, provides a new view of the situation that makes the retreat to Hilbert space unnecessary. The offer wave for each particle can be considered as the wave function of a free (i.e., uncorrelated) particle and can be viewed as existing in normal three-dimensional space. The application of conservation laws and the influence of the variables of the other particles of the system on the particle of interest come not in the offer wave stage of the process but in the formation of the transactions. The transactions "knit together" the various otherwise independent particle wave functions that span a wide range of possible parameter values into a consistent ensemble, and only those wave function sub-components that are correlated to satisfy the conservation law boundary conditions at the transaction vertices are permitted to participate in this transaction formation. The "allowed zones" of Hilbert space arise from the action of transaction formation, not from constraints on the initial offer waves, i.e., particle wave functions.

Thus, the assertion that the quantum wave functions of individual particles in a multi-particle quantum system cannot exist in ordinary three-dimensional space is a misinterpretation of the role of Hilbert space, the application of conservation laws, and the origins of entanglement. It confuses the "map" with the "territory". Offer waves are somewhat ephemeral three-dimensional space objects, but only those components of the offer wave that satisfy conservation laws and entanglement criteria are permitted to be projected into the final transaction, which also exists in three-dimensional space.

These questions are Nick Herbert's "Quantum Mysteries", taken from his online blog "Quantum Tantra: Investigating New Doorways into Nature". It is available online at http://quantumtantra.blogspot.com.

**Q**: *Quantum Mystery # 1a*: What does the quantum wave function really represent?

**A**: The wave function represents an offer to form a transaction that constitutes an interaction and an exchange of conserved quantities. It is a retarded wave that propagates at the natural velocity of the particle it describes through normal three-dimensional space, but it is not, in itself, observable; only the transaction it may form is observable.

**Q**: *Quantum Mystery # 1b*: What is really happening in the world before any measurements are made?

**A**: Retarded offer waves propagate from the source to potential sites for transaction formation, advanced confirmation waves propagate back to the source, the source selects from among the advanced/retarded echos it receives, and a transaction (an advanced/retarded standing wave), possibly one involving a measurement, may form. The transaction is what is "really happening".

**Q**: *Quantum Mystery # 2a*: What really happens during a quantum measurement?

**A**: A quantum measurement is the particular type of transaction that enables an observer to gain information about a quantum system under study.

**Q**: *Quantum Mystery # 2b*: How does a quantum possibility decide to turn into an actuality?

**A**: The offer wave represents a "quantum possibility". It travels to potential sites of transaction formation, where it stimulates them to generate advanced confirmation waves that travel back to the source. The source randomly and hierarchly selects one confirmation for transaction formation, and a transaction forms, turning possibility into actuality.

**Q**: *Quantum Mystery # 3a*: What (if anything) is actually exchanged between two distant entangled quantum subsystems?

**A**: Nothing is directly exchanged between distant entangled quantum subsystems. Each subsystem exchanges retarded and advanced waves with the system producing the entangled subsystems, and the boundary conditions at each transaction vertex enforce the conservation laws that are the basis of the entanglement and the resulting correlation.

**Q**: *Quantum Mystery # 3b*: When will physicists get smart enough to be able to tell their kids a believable story about what's really going on between Alice and Bob?

**A**: After they read this book and understand the Transactional Interpretation.

## A.4 Questions from Tammaro

These are questions taken from Tammaro, reference [23], Sect. 9, which criticizes the Transactional Interpretation and a number of other QM interpretations. Many of the questions about the TI that Tammaro raises are answered in Chaps. 5 and 6, but here are a few that perhaps require special attention.

**Q**: How can the probability relation used for the Dirac Equation, $P = \psi\psi^{\dagger}$, be reconciled with the Transactional Interpretation?

**A**: For the relativistic Dirac wave equation, which is used for fractional-spin fermion particles subject to Fermi-Dirac statistics, the probability is $P = \psi\psi^{\dagger}$, where † indicates that the conjugate-transpose of a column matrix must be taken and a matrix multiplication performed. Is this different from the Born rule discussed above?

No. The use of column and row matrices is a book-keeping technique used in Dirac algebra to keep track of four-momentum components, spin projections, etc., and the transpose operation simply ensures that appropriate pairs of elements in the matrices will be properly paired with and multiplied by their complex conjugates, as the Born rule indicates. The Dirac probability procedure is just the Born rule, generalized for Dirac algebra.

**Q**: How can the probability relation used for the Klein–Gordon Equation, $P = \frac{i\hbar}{2mc^2}(\psi^* \frac{d}{dt}\psi - \psi\frac{d}{dt}\psi^*)$, be reconciled with the Transactional Interpretation and its use of the Born probability rule?

**A**: The problem with the Klein–Gordon (K-G) equation (if it is a problem) is that it has both retarded and advanced wave functions as solutions, and the latter have energy eigenvalues that are negative. The result is that the Born probability $\psi\psi^*$ as applied to general Klein–Gordon wave functions does not satisfy the continuity equation, because it associates positive probabilities with the negative energy solutions. The probability needs to be negative for advanced waves that when emitted carry negative energy backwards in time to a previous point where they may form a transaction by handshaking with a retarded wave. The probability needs to be negative because energy disappears at the "emission" point and moves in the negative time direction, while $\psi\psi^*$ always gives a positive probability value.

As stated above, the K-G probability relation $P = \frac{i\hbar}{2mc^2}(\psi^* \frac{d}{dt}\psi - \psi\frac{d}{dt}\psi^*)$ must be used in the general case. Referring to Appendix B.2, the energy operator that extracts the energy from a wave function is $\mathbb{E} = i\hbar\frac{d}{dt}$, i.e., the time derivative multiplied by $i\hbar$. Therefore, the expression $P = \frac{i\hbar}{2mc^2}(\psi^* \frac{d}{dt}\psi - \psi\frac{d}{dt}\psi^*)$ is equivalent to $(\psi^*\mathbb{E}\psi - \psi\mathbb{E}\psi^*)/(2mc^2)$. The energy operator $\mathbb{E}$ extracts the energy from the wave function or its complex conjugate, subtracts them (which is equivalent to adding, since they will have opposite signs), divides the added eigen-energies by $2mc^2$, and leaves behind the $\psi\psi^*$ Born product. Since both retarded waves with positive energy eigenvalues and advanced waves with negative energy eigenvalues are solutions to the K-G equation, both kinds of waves can be the represented by $\psi$ in the probability equation, and the latter will have negative energy eigenvalues. If $\psi$ is a retarded wave, i.e., contains $\exp(-iEt/\hbar)$, the energy eigenvalue $E = mc^2$ will be positive, so the procedure will give a net positive probability. If $\psi$ is an advanced wave, i.e., contains $\exp(+iEt/\hbar)$, the energy eigenvalue $E = -mc^2$ will be negative, so the procedure will give a net negative probability. Thus the K-G probability expression is just a generalization of the Born rule that includes a book-keeping mechanism for insuring that the probability for both advanced and retarded wave solutions to the Klein–Gordon equation has the proper sign. It is completely consistent with the Transactional Interpretation.

We note that some textbooks attempt to associate the negative-energy K-G solutions with antimatter rather than with advanced waves. This perhaps works for the special case of fermion-composite particles like $\pi$-mesons, but the Klein–Gordon equation describes particles obeying Bose–Einstein statistics or bosons, and fundamental point-like bosons do not have antimatter twins. It is the fermion particles, described by the Dirac equation and obeying Fermi-Dirac statistics and the Pauli principle, that have the peculiarities of half-integer spins and antimatter twins with opposite parity. Further, the association of K-G wave function solutions with "charge currents", i.e., multiplying probabilities by particle charge to eliminate negative probabilities, is questionable because important point-like bosons, e.g., photons, gluons, $Z^0$ bosons, Higgs bosons, and perhaps axions, are electrically uncharged.[5]

**Q**: The Transactional Interpretation fails to deal with the "measurement problem", doesn't it?

**A**: The measurement problem arises from the mind-set of the knowledge interpretation, in that changes in observer knowledge are produced by measurements, but "measurement" has no precise definition. For the TI, a measurement is just another transaction, and the TI avoids the thorny questions of just what a measurement is and when and how it collapses the wave function. The TI does this by means of the atemporal description of transaction formation. See the discussion of the Schrödinger's Cat Paradox in Sect. 6.3 above. Also, see Chap. 3.3 of Kastner's TI book [16] for more detailed analysis of the TI solution to the measurement problem and Carver Mead's mathematical description of transaction formation [15].

## A.5 Questions from "Interpretations of QM" in the *Internet Encyclopedia of Philosophy*

These are questions taken from the "Interpretations of Quantum Mechanics" article by Peter J. Lewis published in the online *Internet Encyclopedia of Philosophy* [27]. The article, despite its general title, seems to be a puff-piece promoting the Everett-Wheeler Many-Worlds Interpretation.

[5] As a practical matter, I have had much experience doing calculations in ultra-relativistic heavy-ion physics in which we numerically solved the Klein–Gordon equation and then used the numerical wave function solutions produced to predict the behavior and correlations of $\pi$ mesons [24–26] from 200 GeV/nucleon ion-ion collisions. In such calculations, we did *not* use the K-G probability relation quoted in the question above. Instead, we calculated the retarded-wave positive-energy K-G wave-function solutions and used these with the $\psi\psi^*$ Born relation to evaluate the probabilities needed in the calculations. This works very well, as long as one does not use the advanced K-G solutions with negative energies, which are easy to recognize and avoid.

**Q**: In the Transactional Interpretation, does the wave function exist in a pre-collapse form?

**A**: The question begs for a definition of "exists". The wave function or offer wave exists as a representation of possibility and as a component of the final process, but it is not in a form that can be experimentally tested to verify its existence. It also exists as a mathematical solution of the appropriate wave equation that satisfies the system-imposed boundary conditions.

**Q**: Does the story involving forwards and backwards waves constitute a genuine explanation of transaction formation?

**A**: The question begs for a definition of "genuine". What is required for a plausible description and explanation (that is visible in the quantum formalism itself) to become a "genuine explanation"? If mathematics is needed, in Sect. 5.6 we discuss Carver Mead's mixed-state calculation [15] using the standard quantum wave mechanics formalism. That provides a mathematical description of the "quantum-jump" transfer of a photon between atoms, a transaction that forms through the exchange of advanced and retarded waves between the atoms.

**Q**: Is the Transactional Interpretation a hidden variable theory?

**A**: Of course not. Complementary variables that form Fourier pairs cannot have simultaneous precise values, either in the standard quantum formalism or in the Transactional Interpretation view of that formalism. See the discussion in Sects. 2.4 and 2.5.

**Q**: Is the Transactional Interpretation a description of the quantum world or an "instrumentalist recipe"?

**A**: We are unsure what an "instrumentalist recipe" is, but it sounds very much like a philosopher's insult to experimental physicists and experimental physics. The Transactional Interpretation is indeed a description of the quantum world, and it is also a recipe for explaining and understanding many otherwise mysterious experiments (see Chap. 6) that the interpretations promoted in Lewis' article appear to be incapable of explaining. See, for example, the "boxed atom" experiments analyzed in Sect. 6.19.

**Q**: Isn't the Everett interpretation the only interpretation that follows directly from a literal reading of the standard theory of quantum mechanics?

**A**: That is a very peculiar assertion in view to the problems that the Everett interpretation has with accommodating the Born Probability rule, entanglement, and nonlocality [28], all parts of standard quantum mechanics. Everett tried to evade the nonlocality issue by calling EPR nonlocality a "false paradox". Further, the Afshar experiment [29] (see Sect. 6.17) demonstrates the experimental presence of interference in a situation in which the Everett interference rule (no interference between distinguishable "worlds") predicts no interference. We would say that the Transactional Interpretation is the only interpretation that follows directly from a literal reading of standard quantum mechanics.

# Appendix B
# A Brief Overview of the Quantum Formalism

Here we will provide a brief summary of the formalism of quantum wave mechanics, keeping at a minimum the need for the reader to have a mathematical background. In this discussion, we will use only one space dimension $x$. The formalism, of course, is normally applied to waves in 3D space, but this requires the use of vectors and gets in the way of explaining what is going on, so we will stick to 1D.

## B.1 Waves Properties and Wave Functions

We will start with the concept of traveling waves. A useful example is a wave/kink traveling down a clothes line. If you hit a taut clothes line with a stick, a downward "kink" travels down the line until it hits the end, where it reflects and an upward kink comes back. If you arrange to wiggle the end of the clothes line up and down continuously and smoothly, ripples in the shape of sine waves travel down the line. Ripples in a pond, ocean waves, earthquake waves, sound waves, and light waves all behave in this same way. The wave has a fairly constant shape, but that shape moves along in the medium (e.g., the clothes line, etc.) with some velocity.

We can focus on just those traveling waves that are sinusoidal (i. e., have the shape of a sine wave). If you take a snapshot of a sinusoidal traveling wave on a clothes line, you can identify its wavelength $\lambda$ as the distance from one maximum to the next. If you observe at one point along its travel, that section of rope will rise and fall and you can identify its period $T$ as the time it takes, starting from a height maximum, to fall to a minimum height and then rise to the next maximum. You can also identify its amplitude $A$ as half the distance between a maximum point and a minimum point as it oscillates. We can use these quantities to define some additional properties of the wave. Its *wave number* $k$ is given by $k = 2\pi/\lambda$ and its *angular frequency* $\omega$ is given by $\omega = 2\pi/T$. The speed $v$ of the wave as it travels in the medium is given by $v = \lambda/T = \omega/k$.



J.G. Cramer, *The Quantum Handshake*, DOI 10.1007/978-3-319-24642-0

Using this information, we can write the *wave function* of the traveling wave as:

$$\psi(x,t) = A\cos(kx - \omega t) = \Re\{A\exp[i(kx - \omega t)]\}, \tag{B.1}$$

where $x$ is a position in space, $t$ is the time, and $\Re\{\}$ means the real part of a complex function (see Chap. 4). The $\omega t$ term here has a minus sign in order to make the wave move along the $+x$ axis, because at later and later times it requires going to a larger and larger $x$-value to keep the argument of the cosine function at the same value, e.g., at a maximum or minimum.

The wave functions of quantum mechanics differ in only one way from the wave functions for sound waves or ripples on a clothes line: we must use the entire complex function instead of using just the real part:

$$\psi(x,t) = A\exp[i(kx - \omega t)]. \tag{B.2}$$

In the quantum world, we will use wave functions for describing massless photons of light, and we will also use wave functions for describing two classes of massive particles, bosons and fermions. For the purposes of this discussion, all of these particle types share the same form for their wave functions, but the quantum wave equations to which these wave functions are solutions may be very different.

## B.2 Wave Function Localization in *x* and *p*

The wave function $\psi(x,t) = A\exp[i(kx - \omega t)]$ of Eq. B.2 is what is called a "plane wave". It assumes that the amplitude $A$ is a constant and describes traveling waves with a definite momentum $p = \hbar k$ and energy $E = \hbar\omega$, but with no definite locations in time or space. However, what do we do if we wish to describe a "wave packet" that is localized in space? The answer is that we must make the amplitude $A$ a function of position $x$, which has the effect of making $k$ a variable with a distribution of values. This makes the wave more likely to be in the place where $A$ is large than in the place where $A$ is small.

If we wish to localize a particle at a position $x_0$ within a position uncertainty of $\sigma_x$ we can write the wave function as

$$\psi(x,t) = (\sigma_x\sqrt{2\pi})^{-1}\exp[-(x - x_0)^2/2\sigma_x^2]\exp[i(kx - \omega t)] \tag{B.3}$$

This makes the amplitude $A$ a unit-area Gaussian distribution centered about $x_0$ with a variance of $\sigma_x$, as shown in Fig. B.1.

What is the effect of this localization on the variable $k$ which is complementary to $x$? We can answer this question, as discussed in Sect. 2.4, by calculating $\phi(k)$, the momentum-space wave function of $\psi(x)$, which is the Fourier transform of $\psi(x)$. It is well known that the Fourier transform of a Gaussian distribution is another Gaussian distribution, with the widths of the two Gaussians in a see-saw relation: The Fourier

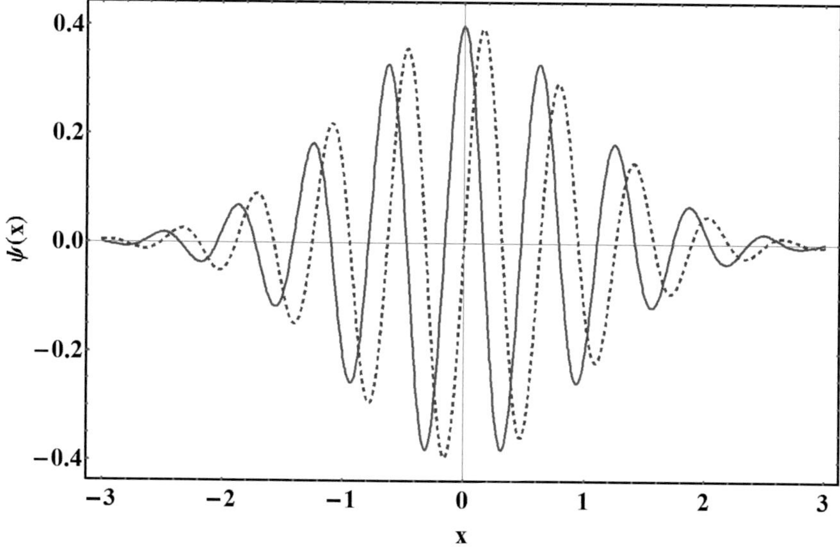


**Fig. B.1** A quantum wave function $\psi(x)$ of Eq. B.3 localized at $x_0 = 0$ with a width $\sigma_x = 1$ and $k = 10$. The real part of $\psi(x)$ is shown as *red/solid* and the imaginary part as *blue/dashed*

transform of $\psi(x)$ gives $\phi(k)$ with a localization uncertainty of $\sigma_k = 1/\sigma_x$. In other words, $\sigma_k\sigma_x = 1$. We can then bring the momentum $p$ into the picture by recalling that $p = \hbar k$. Therefore, in terms of the momentum uncertainty $\sigma_p$, the same relation can be written as $\sigma_p\sigma_x = \hbar$, which is a version of Heisenberg's uncertainty relation (see Sect. 2.4).

## B.3 Operators and Eigenvalues

Now that we have a wave function, we can ask what it describes and how we can access the information that it contains. Let us start by focusing on light waves. Light is a combination of a sinusoidal electric field and a sinusoidal magnetic field, vibrating in phase but at right angles to each other and moving at the speed of light in the direction perpendicular to both fields. Figure B.2 shows a vertically polarized classical light wave, as predicted by Maxwell's equations.

The quantum mechanical version of light differs from this picture because the energy and momentum are *quantized* into photons. Each photon carries an energy of $E = \hbar\omega$ and a momentum of $p = \hbar k$, where $\hbar$ is Planck's constant ($6.62606957 \times 10^{-34}$ m$^2$kg/s) divided by $2\pi$. By substituting in the energy and momentum expressions, we can write the wave function $\psi_\gamma$ that is the offer wave for a photon as: $\psi_\gamma(x, t) = A\exp[\frac{i}{\hbar}(px - Et)]$.

We can see that this wave function contains the photon's energy and momentum as interior variables, but how can we get them out? The trick is to use calculus. The

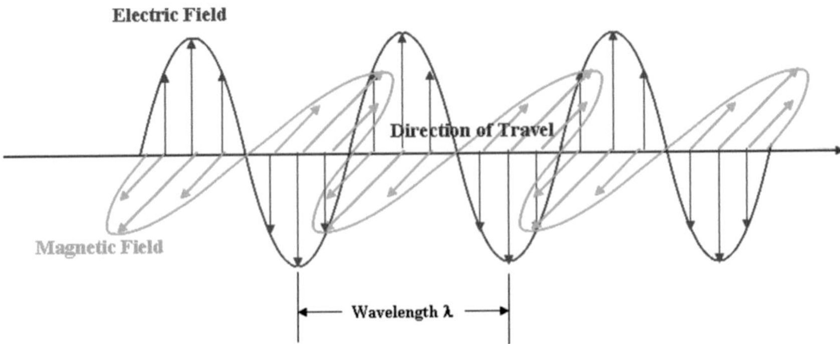


**Fig. B.2** A vertically polarized classical light wave showing the electric (*red*) and magnetic (*green*) field directions, the wavelength $\lambda$, and the direction of wave motion (*pink*)

calculus operation of *differentiation*, $\frac{d}{dx} f(x)$, determines the rate at which a function $f(x)$ changes when $x$ is changed. For example, if $f(x) = x^2$, then $\frac{d}{dx} f(x) = 2x$. So what happens if we differentiate the function $f(x) = \exp(ax)$? The answer is: $\frac{d}{dx}\exp(ax) = a\exp(ax)$. In other words, differentiating the exponential with respect to $x$ leaves it unchanged, except that it pulls out the quantity that is multiplying $x$ inside the exponential.

Therefore, $\frac{d}{dx}\psi_\gamma(x,t) = \frac{ip}{\hbar}\psi_\gamma(x,t)$. In other words, the operation of differentiating the photon wave function with respect to $x$ pulls the quantity $ip/\hbar$ out of the wave function. This is good, but we would rather not have the $i/\hbar$ as extra baggage. Therefore, we can define the *momentum operator*:

$$\mathbb{P} \equiv \frac{\hbar}{i}\frac{d}{dx} \tag{B.4}$$

$$\mathbb{P}\psi_\gamma(x,t) = p\psi_\gamma(x,t). \tag{B.5}$$

The momentum operator $\mathbb{P}$ extracts just the momentum $p$ from the wave function. Similarly, differentiation with respect to the time $t$ will extract a quantity that includes the energy $E$, the variable in the wave function that $t$ multiplies. In particular, we can define the energy operator as $\mathbb{E} \equiv i\hbar\frac{d}{dt}$. Here the $i$ has moved from denominator to numerator because the $Et$ term has a minus sign, and $i = -1/i$. Thus, $\mathbb{E}\psi_\gamma(x,t) = E\psi_\gamma(x,t)$ and energy operator extracts the photon energy $E$ from the wave function. In the language of quantum mechanics, we say that $p$ and $E$ are the momentum and energy *eigenvalues* of the wave function.

If these operators operate *twice* on the wave function, they extract the *square* of the eigenvalue. This allows us to define square operators that extract the momentum-squared and the energy-squared. The momentum-squared operator is $\mathbb{P}^2 \equiv \mathbb{P}\mathbb{P} \equiv -\hbar^2\frac{d^2}{dx^2}$, so that $\mathbb{P}^2\psi_\gamma(x,t) = p^2\psi_\gamma(x,t)$. Similarly, the energy-squared operator is $\mathbb{E}^2 \equiv \mathbb{E}\mathbb{E} \equiv -\hbar^2\frac{d^2}{dt^2}$, so that $\mathbb{E}^2\psi_\gamma(x,t) = E^2\psi_\gamma(x,t)$.

## B.4 Wave Equations and Solutions

So far, we have used the wave functions without asking where they came from. Now we will drop the other shoe. The *wave equation* is a differential equation that relates the wave function to its derivatives, and the wave function is a solution of this wave equation, i.e., a mathematical function that, when differentiated once or twice and substituted into the wave equation will satisfy the indicated equality.

The electromagnetic wave equation is used in both classical mechanics and quantum mechanics. It is derived by manipulating Maxwell's equations, the set of equations that describe the classical behavior of electric and magnetic fields. The electromagnetic wave equation for the electromagnetic vector potential $A$ of an electromagnetic wave is the differential equation $\frac{d^2}{dt^2}A = c^2\frac{d^2}{dx^2}A$. The quantum mechanical version of the electromagnetic wave equation is:

$$\frac{d^2}{dt^2}\psi_\gamma = \frac{\omega^2}{k^2}\frac{d^2}{dx^2}\psi_\gamma = c^2\frac{d^2}{dx^2}\psi_\gamma. \tag{B.6}$$

We note that the operations performed on $\psi_\gamma$ are similar to the momentum-squared operator $\mathbb{P}^2$ and energy-squared operator $\mathbb{E}^2$ defined in Sect. B.2 above. In particular, using these operators, we can rewrite the wave equation as:

$$(-\mathbb{E}^2/\hbar^2)\psi_\gamma = c^2(-\mathbb{P}^2/\hbar^2)\psi_\gamma \tag{B.7}$$

or more simply as:

$$\mathbb{E}^2\psi_\gamma = c^2\mathbb{P}^2\psi_\gamma. \tag{B.8}$$

Thus, in operator terms the electromagnetic wave equation is just the operator version of the energy-momentum relation of a photon, i.e., $E = cp$ or $E^2 = c^2p^2$.

The energy-momentum relation for massive particles is different from the above photon relation. In particular, if $E$ is the kinetic energy and $p$ the momentum of a particle of mass $m$, then the non-relativistic relation between them is $p^2/2m = E$. This suggests that, using operators, the wave equation for a massive particle should be similar, i.e., that:

$$(1/2m)\mathbb{P}^2\psi_m = \mathbb{E}\psi_m. \tag{B.9}$$

Rewriting this using the above definitions of $\mathbb{P}^2$ and $\mathbb{E}$ gives:

$$\frac{-\hbar^2}{2m}\frac{d^2}{dx^2}\psi_m = i\hbar\frac{d}{dt}\psi_m \tag{B.10}$$

or more simply;

$$\frac{i\hbar}{2m}\frac{d^2}{dx^2}\psi_m = \frac{d}{dt}\psi_m. \tag{B.11}$$

This is the Schrödinger wave equation, the fundamental non-relativistic wave equation of quantum mechanics.

Thus, by the straightforward use of energy-momentum relations and operators we have derived that fundamental wave equations of quantum mechanics. The quantum mechanical wave function given in Sect. B.1 above is the solution for both the electromagnetic wave equation and the Schrödinger equation.

In the domain of relativistic quantum mechanics, the Klein–Gordon equation that is the appropriate wave equation for boson particles can be derived from the relativistic energy-momentum relation $E^2 = (pc)^2 + (mc^2)^2$, where $E$ is now the *total* energy of the system (including the rest mass-energy). The Dirac equation, appropriate for fermions, can be similarly derived from relativistic energy-momentum relations, but in a somewhat more complicated way because it includes spin vectors.

## B.5 Making Predictions

The use of operators to extract values of variables from the wave function, as described above, is incomplete, because the extracted quantity is multiplied by the wave function. The question of how to make a prediction of the outcome of a measurement remains to be defined.

As indicated in Sect. B.2, we can construct operators that can extract quantities like energy and momentum from the wave function. These can be used to calculate a prediction of the "expectation value" of the quantity by making a "quantum mechanical sandwich" with wave functions as the "bread" and an operator as the "meat". This is interpreted by the TI as follows: The offer wave $\psi$ contains information that can be extracted by some operator $\mathbb{Q}$ (perhaps a device that measures the quantity $q$) so that $\mathbb{Q}\psi = q\psi$. Multiplying this by the confirmation wave $\psi^*$ gives $\psi^*(\mathbb{Q}\psi) = q\psi\psi^* = (prob)q$. The latter is the probability of the variable of interest having a value of $q$ at the location of the wave function $\psi$. Since we are interested in the overall value, this must be averaged over all space to get the "expectation value" of $q$, which we represent as $\langle q \rangle$.

For example, suppose that we want to calculate the expectation value $\langle E \rangle$ of the energy of a system described by wave function $\psi(x)$. The procedure to calculate it is:

$$\langle E \rangle = \int_{-\infty}^{\infty} \psi^*(x)\mathbb{E}\psi(x)dx. \tag{B.12}$$

This works because $\psi^*\mathbb{E}\psi = \psi^* E \psi = \psi^* \psi E = (\psi\psi^*)E = PE$, where $P$ is the probability of finding the quantity E at a particular space-time location. The integration adds up all of the probabilities, effectively averaging E over all space.

Similarly, if we wanted the expectation value $\langle p \rangle$ of the momentum, we would calculate:

$$\langle p \rangle = \int_{-\infty}^{\infty} \psi^*(x)\mathbb{P}\psi(x)dx. \tag{B.13}$$

That's how quantum wave mechanics calculations are done. The Transactional Interpretation provides insights into why this mathematics describes Nature in action.

# Appendix C
# Quantum Dice and Poker—Nonlocal Games of Chance

In this Appendix we want to describe two games of chance that are analogous to the behavior of entangled particles. These may help in understanding how non-classical the quantum game is.

## C.1 Quantum Dice

Suppose that you are given two dice that have somehow been entangled by a conservation law that requires the number seven. You roll one of the dice and get a random face between one and six. You can plot the results of many such rolls, and you find a completely flat distribution, which each face equally probable.

Then you roll its entangled twin, and the face that appears, when added to that from the first roll, is always seven. However, if you plot the distribution of faces from many rolls, again there is a completely flat distribution. The dice are, at the same time, completely random in the numbers they produce individually, and perfectly correlated in their sum, in that together they always produce a seven.

In this way, the quantum dice behave in the same way as a pair of entangled particles, which give random but correlated results of selected measurements. Perhaps, contrary to Einstein's expectations, the Old One does play dice, and they are loaded by entanglement.

## C.2 The Rules of Quantum Poker

Quantum poker is a "game" played with ordinary playing cards that illustrates the nature of the nonlocal quantum correlations in EPR situations. Alice and Bob are playing a game of quantum poker. The deck, handled only by the Dealer, consists of eight cards, two copies each of four card types: $[A\spadesuit]$, $[K\spadesuit]$, $[A\heartsuit]$, and $[K\heartsuit]$.

 
J.G. Cramer, *The Quantum Handshake*, DOI 10.1007/978-3-319-24642-0

Each players chooses a "measurement" to make during the game and whispers to the Dealer their choice of measurement. The measurement selected may be either *color* (red or black) or *hi/lo* (ace or king). The Dealer randomly draws one card from the deck, selects its duplicate from the deck, and gives one of these cards to Alice and one to Bob. The Dealer then gives each player a second card, based on their measurement choice. If the player is measuring *color*, the Dealer gives the player the other card in the deck that is the same color; if the player is measuring *hi/lo*, the Dealer gives the player the other high or low card from the deck.

*Examples*: (A1) Alice declares that she is measuring color, receives the [$A\heartsuit$], and the Dealer also gives her the [$K\heartsuit$]; (A2) Alice is measuring hi/lo, receives the [$A\heartsuit$], and the Dealer also gives her the [$A\spadesuit$]; (B1) Bob is measuring hi/lo, receives the [$K\heartsuit$], and the Dealer also gives him the [$K\spadesuit$]; (B2) Bob is measuring color, receives the [$K\heartsuit$], and the Dealer also gives him the [$A\heartsuit$].

Now, by looking at his cards, Bob is asked to determine which measurement-type Alice had selected. Does he have enough information to do that? No! He knows that one of his cards matches one of Alice's, but he does not know which one, and he has no information about Alice's other card, so he cannot tell if his hand matches hers or not. All he sees is a series of pairs that randomly correspond to the two alternative colors or the two alternative hi/lo values he has chosen to measure.

## C.3 The Implications of Quantum Poker

We can see that when Bob choses to do the *same* measurement as Alice, he always receives the same resulting card pair. However, if he choses the other measurement, he always gets a non-matching result. In any case, it should be clear that Alice cannot send a message to Bob by using her measurement choice, and Bob cannot receive such a message.

This game of quantum poker is exactly the same situation as if Alice and Bob were receiving polarization-entangled photons, as in the Freedman-Clauser experiment described in Sect. 6.8. They could, for example, choose to measure either horizontal/vertical linear polarization or diagonal/antidiagonal 45° linear polarization for their received photon. If Alice and Bob choose the same measurement, they will always get the same result. However, if they choose to do different measurements, they will get uncorrelated results. In any case, Alice cannot send a message to Bob using her choice of polarization measurements, even though their photons are entangled.

One question remains: how does Nature arrange to play the role of the Dealer in the card game of quantum entanglement, in which Alice and Bob may be very far apart? The Transactional Interpretation provides an answer to this question, as discussed in Sect. 6.6.

# Appendix D
# Detailed Analyses of Selected *Gedankenexperiments*

## D.1 Hardy One-Atom Experiment (Sect. 6.17.1)

The Transactional Interpretation provides a step-by-step description of any quantum event that involves the transfer of conserved quantities (energy, momentum, particle number, angular momentum, angular momentum projection, …). We will apply the TI to the quantum events of the Hardy single-atom interaction-free measurement scenario, described in Sect. 6.17.1 above and shown here in Fig. D.1. Here two coupled transactions are involved: (1) the emission at $L$ and the absorption at $C$, $D$, or $Z+$ of the single photon, and (2) the emission at $X0$ and the absorption at $X\pm$ of the single atom. As in Sect. 6.12, we will explicitly indicate the offer waves by specification of the path as a Dirac ket state vector $\mid path\rangle$, underlining the symbols corresponding to optical elements at which a reflection has occurred. Similarly, confirmation waves will be indicated by a Dirac bra state vector $\langle path\mid$, listing the elements in the time-reversed path with reflections underlined.

First, let us consider the boundary conditions at the six transaction vertices $L$, $C$, $D$, $X0$, $Z+$, and $X\pm$. The photon source $L$ is permitted to emit precisely one photon of the appropriate wavelength during the measurement period. Detectors $C$ and $D$ are permitted to absorb either a single photon in one of the detectors or no photons in either detector. The atom at $Z+$ is permitted to absorb one photon. The initial atom vertex $X0$ must emit exactly one atom, which has an X-axis spin projection of $+{}^1\!/_2$. The final atom vertex $X\pm$ must absorb exactly one atom and report its X-axis spin projection, which may be either $+{}^1\!/_2$ or $-{}^1\!/_2$.

Next, let us consider some particular components of the offer wave initially sent by the photon source $L$. In particular, consider the offer wave components that terminate at detectors $C$ and $D$ and atom $Z+$. These are:



J.G. Cramer, *The Quantum Handshake*, DOI 10.1007/978-3-319-24642-0

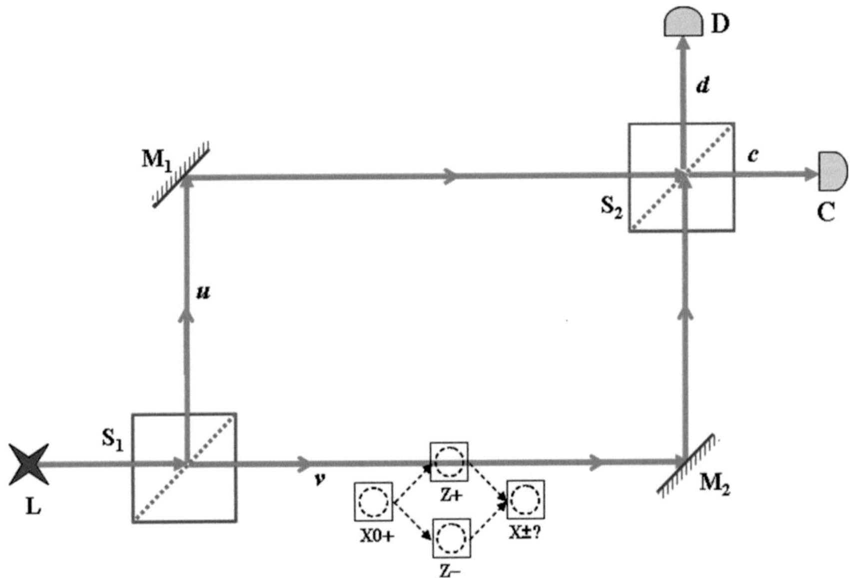


**Fig. D.1** The hardy single-atom interaction-free measurement

$$|\,1\,\rangle \equiv |\,L : \underline{S_1} : \underline{M_1} : S_2 : C\rangle \tag{D.1}$$

$$|\,2\,\rangle \equiv |\,L : \underline{S_1} : \underline{M_1} : \underline{S_2} : D\rangle \tag{D.2}$$

$$|\,3\,\rangle \equiv |\,L : S_1 : \underline{M_2} : \underline{S_2} : C\rangle \tag{D.3}$$

$$|\,4\,\rangle \equiv |\,L : S_1 : \underline{M_2} : S_2 : D\rangle \tag{D.4}$$

$$|\,5\,\rangle \equiv |\,L : S_1 : Z+\rangle. \tag{D.5}$$

Examples of offer wave paths are shown in Fig. D.2.

The arrival of these offer wave components will stimulate confirmation waves that are returned from detectors $C$ and $D$, and atom $Z+$ and travel back to photon source $L$. These are:

$$\langle\,6\,| \equiv \langle C : S_2 : \underline{M_1} : \underline{S_1} : L\,| \tag{D.6}$$

$$\langle\,7\,| \equiv \langle D : \underline{S_2} : \underline{M_1} : \underline{S_1} : L\,| \tag{D.7}$$

$$\langle\,8\,| \equiv \langle C : \underline{S_2} : \underline{M_2} : S_1 : L\,| \tag{D.8}$$

$$\langle\,9\,| \equiv \langle D : S_2 : \underline{M_2} : S_1 : L\,| \tag{D.9}$$

$$\langle\,10\,| \equiv \langle Z+ : S_1 : L\,|\,. \tag{D.10}$$

There are also "dead end" confirmation waves from $C$ and $D$ to the backside of the atom that will cancel and vanish. These are:

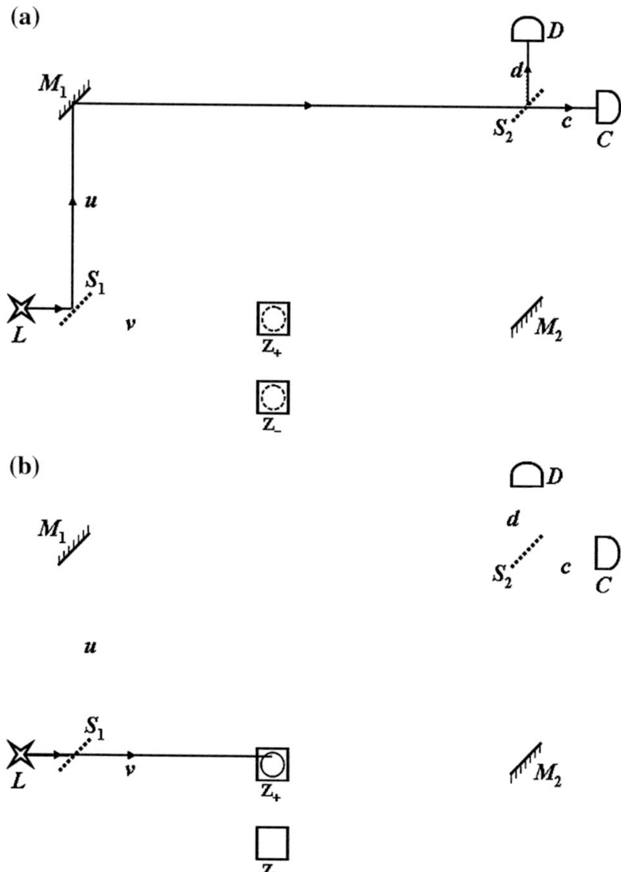


**Fig. D.2** Offer waves **a** | 1 ⟩ and | 2 ⟩ and **b** | 5 ⟩

$$\langle\, 11 \,| \;\equiv\; \langle C : \underline{S_2} : \underline{M_2} : Z_+ \,| \tag{D.11}$$

$$\langle\, 12 \,| \;\equiv\; \langle D : S_2 : \underline{M_2} : Z+ \,| \tag{D.12}$$

The cancellation occurs because they were stimulated by equal-intensity offer waves that were already 90° out of phase because of the reflection to $D$ at $S_2$, and the time-reversed confirmation wave from $C$ receives an additional −90° phase shift on reflection from $S_2$ on its path to the atom. Thus, along the path to the atom the two confirmation waves have equal amplitudes and a 180° phase difference, so that they will cancel. We note that this subtlety was missed in reference [22] and was pointed out in reference [30].

The offer wave components from the atom source $X0$ to the atom detector $X\pm$ are:

$$| \, a \, \rangle \equiv | \, X0_+ : Z+ : X+\rangle \quad \text{(D.13)}$$
$$| \, b \, \rangle \equiv | \, X0_+ : Z+ : X-\rangle \quad \text{(D.14)}$$
$$| \, c \, \rangle \equiv | \, X0_+ : Z- : X+\rangle \quad \text{(D.15)}$$
$$| \, d \, \rangle \equiv | \, X0_+ : Z- : X-\rangle. \quad \text{(D.16)}$$

These offer waves will stimulate corresponding confirmation waves to $X0$, which are:

$$\langle \, e \, | \equiv \langle X+ : Z+ : X0_+ \, | \quad \text{(D.17)}$$
$$\langle \, f \, | \equiv \langle X- : Z+ : X0_+ \, | \quad \text{(D.18)}$$
$$\langle \, g \, | \equiv \langle X+ : Z- : X0_+ \, | \quad \text{(D.19)}$$
$$\langle \, h \, | \equiv \langle X-; Z- : X0_+ \, | \, . \quad \text{(D.20)}$$

If there had been no interaction between the photon and the atom (or if the $Z+$ box had not been in the photon's path), the photon offer waves $| \, 2 \, \rangle$ and $| \, 4 \, \rangle$ and atom offer waves $| \, b \, \rangle$ and $| \, d \, \rangle$ would be in destructive interference superpositions and would cancel. This prevents detection of a photon at "dark" detector $D$ and prevents measurement of X-axis $-{}^1\!/_2$ spin projection for the atom. The photon-atom interaction with a 100 % interaction strength, as assumed by Hardy, means that photon offer waves $| \, 3 \, \rangle$ and $| \, 4 \, \rangle$ may be completely blocked by the presence of the atom in the $Z+$ box. The atom offer waves $| \, c \, \rangle$ and $| \, d \, \rangle$ will also be absent because absorption of the photon will place that atom in an excited state in box $Z+$, which breaks the coherent superposition of the two Z-axis projections. The source $L$ has one photon to emit and three confirmation types to choose from: (1) $\langle \, 10 \, |$: no photon is detected at $C$ or $D$ because the photon was absorbed at the $Z+$ box, and the subsequent X-axis spin measurement on the atom had equal probabilities for $X$-axis spin $+{}^1\!/_2$ and $-{}^1\!/_2$, because the state superposition was broken by the atomic excitation; (2) $\langle \, 6 \, |$ and $\langle \, 8 \, |$: the photon is detected at $C$ and the subsequent X-axis spin measurement on the atom showed a preference for $X$-axis spin $+{}^1\!/_2$, with some probability of $-{}^1\!/_2$; and (3) $\langle \, 7 \, |$: the photon is detected at $D$, the atom is present in box $Z+$ suppressing $\langle \, 9 \, |$, and the subsequent X-axis spin measurement on the atom had equal probabilities for $X$-axis spin $+{}^1\!/_2$ and $-{}^1\!/_2$, i.e., transactions $\langle a \mid e\rangle$ and $\langle b \mid f\rangle$.

Hardy focused on outcome (3), and so we will examine it in detail. The photon is detected at $D$ because the destructive superposition of offer waves $| \, 2 \, \rangle$ and $| \, 4 \, \rangle$ has been broken because offer wave $| \, 4 \, \rangle$ is blocked by the atom, which must be in box $Z+$. The photon transaction is confirmed only by $\langle \, 7 \, |$, because $\langle \, 9 \, |$ is also blocked by the atom. If this transaction forms, no other transaction may form because of the one-photon boundary condition at source $L$. Similarly, the atom can have either of two transactions, offer wave $| \, a \, \rangle$ or $| \, b \, \rangle$ confirmed by confirmation waves $| \, e \, \rangle$ or $| \, f \, \rangle$, respectively, so that measurement of X-axis spin projections $+{}^1\!/_2$ and $-{}^1\!/_2$ are equally probable. However, only one of these transactions may form, because of the one-atom boundary condition at $X0$.

Thus, photon detection at $D$ means that the atom is definitely in box $Z+$ and that the atom, initially prepared at $X0$ in X-axis spin state $+1/2$, can be found with equal probability at $X$ in either spin state, even though there has been no real interaction between the photon and the atom. The Transactional Interpretation provides a way of visualizing this non-classical result and understanding its origin.

## D.2 Polarization-Entangled EPR Experiment (Sect. 7.3)

Figure D.3 shows a polarization-entangled EPR experiment much like the Freedman-Clauser expeeriment, except that we are using a variable-entanglement source like the one described in Sect. 7.3. We initially set $\alpha = 0$ for 100 % entanglement. When $\theta$ is zero and the polarimeters are aligned, there will be a perfect anti-correlation between the polarizations measured by Alice and by Bob. The random polarization (H or V) that Alice measures will always be the opposite of that measured by Bob ($H_A V_B$ or $V_A H_B$). However, when $\theta$ is increased, the perfect $H_A V_B$ and $V_A H_B$ anti-correlations are degraded and correlated detections $H_A H_B$ and $V_A V_B$, previously not present, will begin to appear. Local theories require that for small $\theta$ rotations this correlation degradation should increase linearly with $\theta$, while quantum mechanics predicts that it should increase as $\theta^2$, i.e., quadratically [31]. This is the basis of Bell's Inequalities discussed in Sect. 2.8.

The quantum mechanical analysis of this system is fairly simple because, assuming that the entangled photons have a single spatial mode, their transport through the system can be described by considering only the phase shifts and polarization selections that the system elements create in the waves. We have used the formalism of Horne, Shimony and Zeilinger [32] to perform such an analysis and to calculate the joint wave functions for simultaneous detections at both detectors. These are:

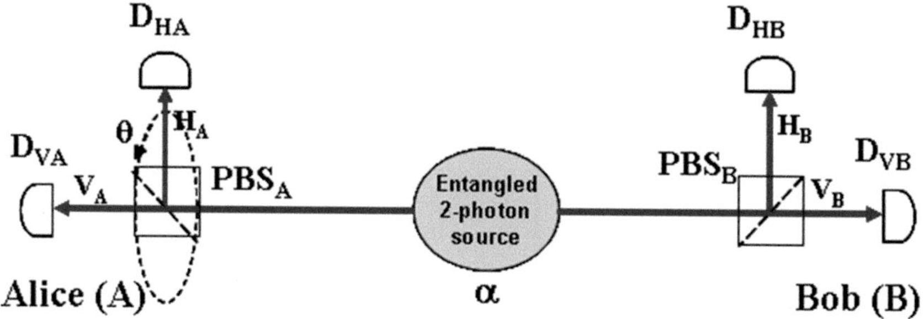


**Fig. D.3** A two-photon 4-detector EPR experiment using linear polarization with variable entanglement

$$\Psi_{HH}(\alpha,\theta) = [-\sin(\alpha)\cos(\theta) + i\cos(\alpha)\sin(\theta)]/\sqrt{2} \quad \text{(D.21)}$$

$$\Psi_{HV}(\alpha,\theta) = [-\cos(\alpha)\cos(\theta) + i\sin(\alpha)\sin(\theta)]/\sqrt{2} \quad \text{(D.22)}$$

$$\Psi_{VH}(\alpha,\theta) = [\cos(\alpha)\cos(\theta) - i\sin(\alpha)\sin(\theta)]/\sqrt{2} \quad \text{(D.23)}$$

$$\Psi_{VV}(\alpha,\theta) = [\sin(\alpha)\cos(\theta) - i\cos(\alpha)\sin(\theta)]/\sqrt{2}. \quad \text{(D.24)}$$

The corresponding joint detection probabilities are:

$$P_{HH}(\alpha,\theta) = [1 - \cos(2\alpha)\cos(2\theta)]/4 \quad \text{(D.25)}$$

$$P_{HV}(\alpha,\theta) = [1 + \cos(2\alpha)\cos(2\theta)]/4 \quad \text{(D.26)}$$

$$P_{VH}(\alpha,\theta) = [1 + \cos(2\alpha)\cos(2\theta)]/4 \quad \text{(D.27)}$$

$$P_{VV}(\alpha,\theta) = [1 - \cos(2\alpha)\cos(2\theta)]/4. \quad \text{(D.28)}$$

Figure D.4 shows plots of these joint detection probabilities versus $\theta$ for the four detector combinations with: $\alpha = 0$ (100 % entangled), $\alpha = \pi/8$ (71 % entangled),

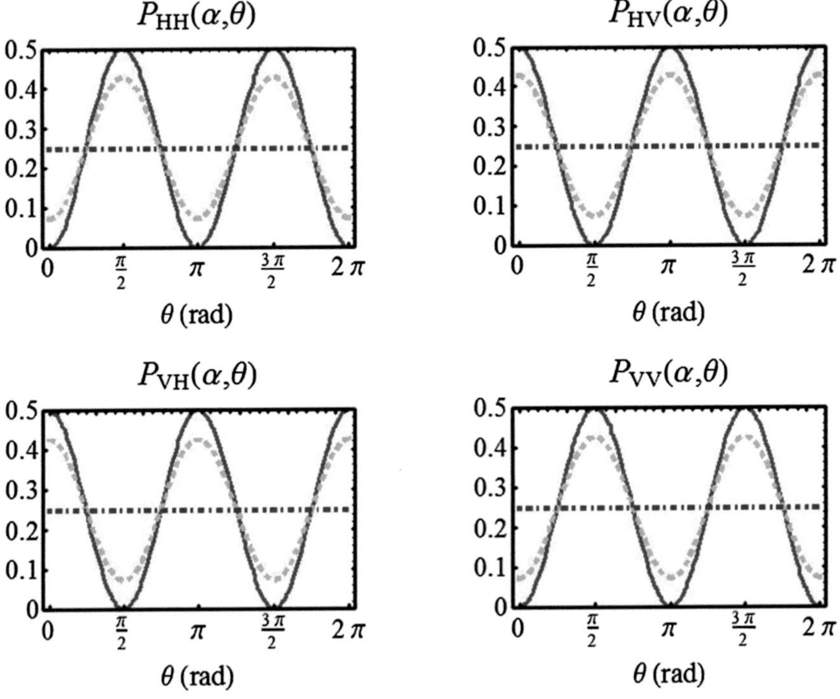


**Fig. D.4** Joint detection probabilities versus $\theta$ for the four detector combinations with: $\alpha = 0$ (*red/solid*, 100 % entangled), $\alpha = \pi/8$ (*green/dashed*, 71 % entangled), and $\alpha = \pi/4$ (*blue/dot-dashed*, 0 % entangled)

and $\alpha = \pi/4$ (0 % entangled). One can see from these plots and equations that the singles probabilities are: $P_{HH}+P_{HV} = P_{VH}+P_{VV} = P_{HH}+P_{VH} = P_{HV}+P_{VV} = 1/2$, and so no signaling is possible.

## D.3 Path-Entangled EPR Experiment* (Sect. 7.4)

As in the polarization-entangled EPR example discussed in Sect. 7.3, the quantum mechanical analysis of this system is fairly simple because, assuming that the entangled photons have a single spatial mode, their transport through the system can be described by considering the phase shifts that the system elements create in the waves. To test the validity of the above arguments, we have used the formalism of Horne, Shimony and Zeilinger [32] to analyze the dual-interferometer configuration shown in Fig. D.5 and to calculate the joint wave functions for detections of the entangled photon pairs in various combinations.

For $BS_A$ in, these wave functions are:

$$\begin{aligned}\Phi_{A_1B_1}(\alpha,\phi_A,\phi_B) &= [i\cos(\alpha)(e^{i\phi_A}-e^{i\phi_B})\\ &\quad+\sin(\alpha)(1+e^{i(\phi_A+\phi_B)})]/(2\sqrt{2}) \qquad (D.29)\\ \Phi_{A_1B_0}(\alpha,\phi_A,\phi_B) &= [-\cos(\alpha)(e^{i\phi_A}+e^{i\phi_B})\\ &\quad+i\sin(\alpha)(1-e^{i(\phi_A+\phi_B)})]/(2\sqrt{2}) \qquad (D.30)\\ \Phi_{A_0B_1}(\alpha,\phi_A,\phi_B) &= [\cos(\alpha)(e^{i\phi_A}+e^{i\phi_B})\\ &\quad+i\sin(\alpha)(1-e^{i(\phi_A+\phi_B)}]/(2\sqrt{2}) \qquad (D.31)\\ \Phi_{A_0B_0}(\alpha,\phi_A,\phi_B) &= [i\cos(\alpha)(e^{i\phi_A}-e^{i\phi_B})\\ &\quad-\sin(\alpha)(1+e^{i(\phi_A+\phi_B)}]/(2\sqrt{2}). \qquad (D.32)\end{aligned}$$

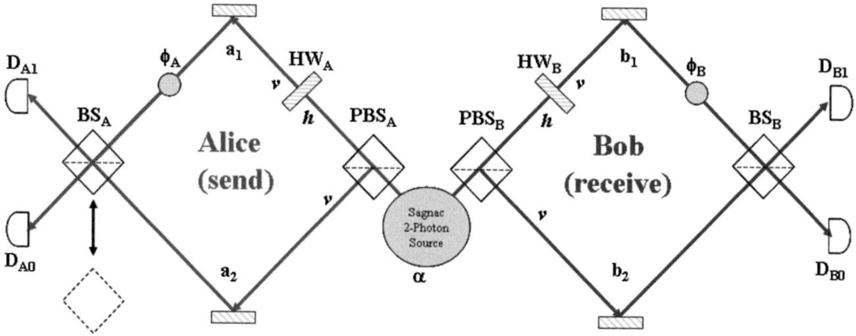


**Fig. D.5** A 4-detector path-entangled dual-interferometer EPR experiment with variable entanglement

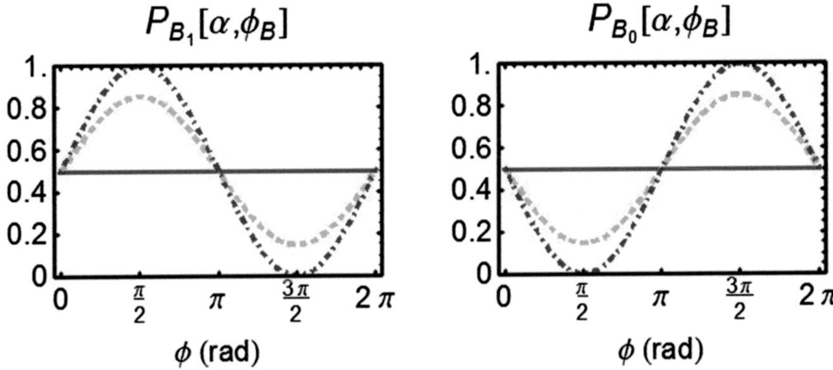


**Fig. D.6** Bob's non-coincident singles detector probabilities $P_{B_1}(\alpha, \phi_B)$ and $P_{B_0}(\alpha, \phi_B)$ (Eqs. D.37 and D.38) for $\alpha = 0$ (*red/solid*, 100 % entangled), $\alpha = \pi/8$ (*green/dash*, 71 % entangled), and $\alpha = \pi/4$ (*blue/dot-dash*, 0 % entangled)

The corresponding joint detection probabilities are:

$$P_{A_1B_1}(\alpha, \phi_A, \phi_B) = \{1 - \sin(\phi_A)[\sin(2\alpha) + \sin(\phi_B)] - \cos(2\alpha)\cos(\phi_A)\cos(\phi_B) + \sin(2\alpha)\sin(\phi_B)\}/4 \tag{D.33}$$

$$P_{A_1B_0}(\alpha, \phi_A, \phi_B) = \{1 - \sin(2\alpha)[(\sin(\phi_A) + \sin(\phi_B)] + \cos(2\alpha)\cos(\phi_A)\cos(\phi_B) + \sin(2\alpha)\sin(\phi_B)\}/4 \tag{D.34}$$

$$P_{A_0B_1}(\alpha, \phi_A, \phi_B) = \{1 + \sin(2\alpha)[(\sin(\phi_A) + \sin(\phi_B)] + \cos(2\alpha)\cos(\phi_A)\cos(\phi_B) + \sin(2\alpha)\sin(\phi_B)\}/4 \tag{D.35}$$

$$P_{A_0B_0}(\alpha, \phi_A, \phi_B) = \{1 - \sin(\phi_B)[\sin(2\alpha) + \sin(\phi_A)] - \cos(2\alpha)\cos(\phi_A)\cos(\phi_B) + \sin(2\alpha)\sin(\phi_A)\}/4. \tag{D.36}$$

The non-coincident singles detector probabilities for Bob's detectors are obtained by summing over Alice's detectors, which he does not observe. Thus

$$P_{B_1}(\alpha, \phi_B) \equiv P_{A_1B_1}(\alpha, \phi_A, \phi_B) + P_{A_0B_1}(\alpha, \phi_A, \phi_B) = [1 + \sin(2\alpha)\sin(\phi_B)]/2 \tag{D.37}$$

$$P_{B_0}(\alpha, \phi_B) \equiv P_{A_1B_0}(\alpha, \phi_A, \phi_B) + P_{A_0B_0}(\alpha, \phi_A, \phi_B) = [1 - \sin(2\alpha)\sin(\phi_B)]/2. \tag{D.38}$$

Note that these singles probabilities have no dependences on Alice's phase $\phi_A$ for any value of $\alpha$. Here again we see an example of Schrödinger steering, in that Alice is manipulating the wave functions that arrive at Bob's detectors, but not in such a way that would permit signaling.

Figure D.6 shows plots of Bob's non-coincident singles detector probabilities $P_{B_1}(\alpha, \phi_B)$ and $P_{B_0}(\alpha, \phi_B)$ for the cases of $\alpha = 0$ (100 % entangled), $\alpha = \pi/8$ (71 % entangled), and $\alpha = \pi/4$ (not entangled).

We see here a demonstration of the see-saw relation between entanglement and coherence [33], in that the probabilities for fully entangled system are constant, independent of $\phi_B$, because the absence of coherence suppresses the Mach–Zehnder interference, while the unentangled system shows strong Mach–Zehnder interference. The $\alpha = \pi/8$ case, with 71 % coherence and entanglement, also shows fairly strong Mach–Zehnder interference and raises the intriguing possibility that a nonlocal signal might survive.

Therefore, the question raised by the possibility of nonlocal signaling is: *What happens to Bob's detection probabilities when Alice's beam splitter $BS_A$ is removed?* To answer this question, we re-analyze the dual interferometer experiment of Fig. D.5 with $BS_A$ in the "out" position. These calculations give the joint wave functions for simultaneous detections of detector pairs:

$$\Psi_{A_1B_1}(\alpha, \phi_A, \phi_B) = [\sin(\alpha) - ie^{i\phi_B}\cos(\alpha)]/2 \tag{D.39}$$

$$\Psi_{A_1B_0}(\alpha, \phi_A, \phi_B) = [i\sin(\alpha) - e^{i\phi_B}\cos(\alpha)]/2 \tag{D.40}$$

$$\Psi_{A_0B_1}(\alpha, \phi_A, \phi_B) = [e^{i\phi_A}(\cos(\alpha) - ie^{i\phi_B}\sin(\alpha)]/2 \tag{D.41}$$

$$\Psi_{A_0B_0}(\alpha, \phi_A, \phi_B) = [ie^{i\phi_A}(\cos(\alpha) + ie^{i\phi_B}\sin(\alpha)]/2. \tag{D.42}$$

The corresponding joint detection probabilities are:

$$P_{A_1B_1}(\alpha, \phi_A, \phi_B) = [1 + \sin(2\alpha)\sin(\phi_B)]/4 \tag{D.43}$$

$$P_{A_1B_0}(\alpha, \phi_A, \phi_B) = [1 - \sin(2\alpha)\sin(\phi_B)]/4 \tag{D.44}$$

$$P_{A_0B_1}(\alpha, \phi_A, \phi_B) = [1 + \sin(2\alpha)\sin(\phi_B)]/4 \tag{D.45}$$

$$P_{A_0B_0}(\alpha, \phi_A, \phi_B) = [1 - \sin(2\alpha)\sin(\phi_B)]/4. \tag{D.46}$$

The non-coincident singles detector probabilities for Bob's detectors are identical to the singles detector probabilities of Eqs. D.37 and D.38 obtained when $BS_A$ was in place.

## D.4 Wedge-modified Path-Entangled EPR Experiment (Sect. 7.5)

We have performed this analysis of the experiment shown in Fig. D.7, tweaking the mirror angles for maximum overlap of the waves on the two paths to detector $D_A$. The calculation gives large analytical expressions for joint detection probability as a function of position on detector $D_A$, but these must be integrated numerically to

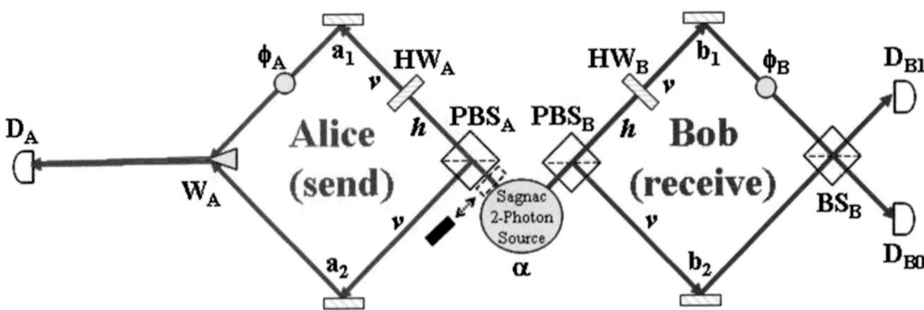


**Fig. D.7** A 3-detector wedge modification of the path-entangled dual-interferometer EPR experiment with variable entanglement

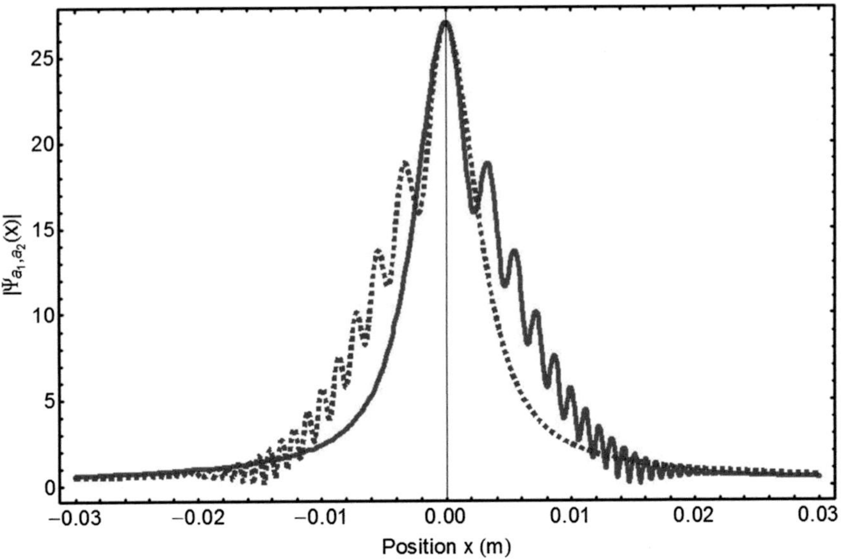


**Fig. D.8** Magnitudes of the wave functions $\Psi_{a1}$ (*red/solid*) and $\Psi_{a2}$ (*blue/dotted*) as functions of position $x$ on the face of detector $D_A$. Oscillations are the result of Gaussian tail truncation by the apex of wedge mirror $W_A$

obtain the position-independent probabilities. Here Fig. D.8 shows the overlap of the magnitudes of the wave functions for paths $a_1$ and $a_2$ versus position. The wave functions have a basic Gaussian profile with oscillations arising from the truncation of one Gaussian tail by $W_A$.

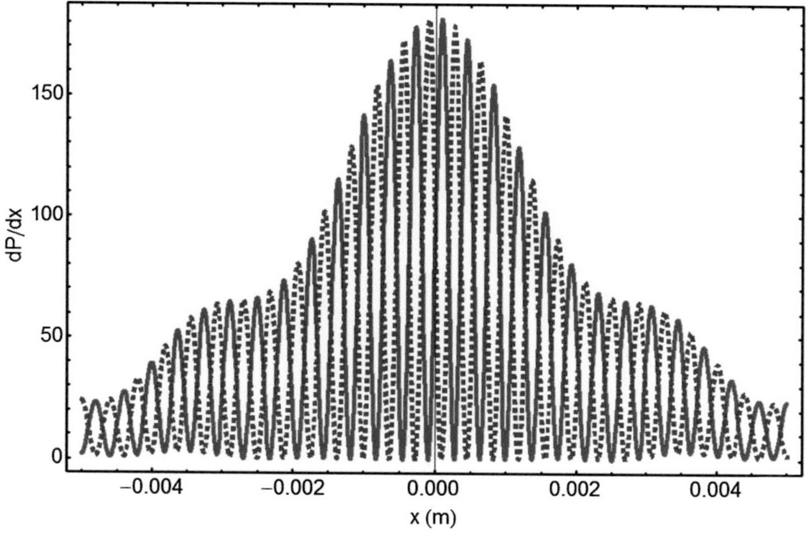


**Fig. D.9** Probabilities of coincident detections at $D_A$ and $D_{B_1}$ (*red/solid*) and at $D_A$ and $D_{B0}$ (*blue/dotted*) with $\alpha = 0$, $\phi_A = 0$, and $\phi_B = 0$

Figure D.10 shows the corresponding probabilities for $\alpha = 0$ (e.g., fully entangled) of coincident photon pairs at Alice's detector $D_A$ and at Bob's detectors $D_{B_1}$ and $D_{B_0}$. The probabilities are highly oscillatory because of the interference of the two waves and the phase walk of the wave functions with angle, analogous to two-slit interference.

To test the possibility of a nonlocal signal, we must integrate these probabilities over the extent of the detector face and calculate difference functions from these results and similar evaluations of Eqs. D.37 and D.38. We can expect some errors in numerical integration due to the oscillation shown in Fig. D.9. The difference functions as 2-D contour plots in $\phi_B$ versus $\alpha$ are shown in Fig. D.10.

Thus, the differences between the probabilities predicted by of Eqs. D.37 and D.38 and the numerically-integrated probabilities of Fig. D.10 are on the order of a few parts per million. Does this mean that there is a small residual nonlocal signal? No! It means that any comparison involving numerical integration is subject to round-off error and is not reliable beyond a few parts per million. The results shown in Fig. D.10 demonstrate that no nonlocal signal is possible using the wedge-modified configuration of Fig. D.7.

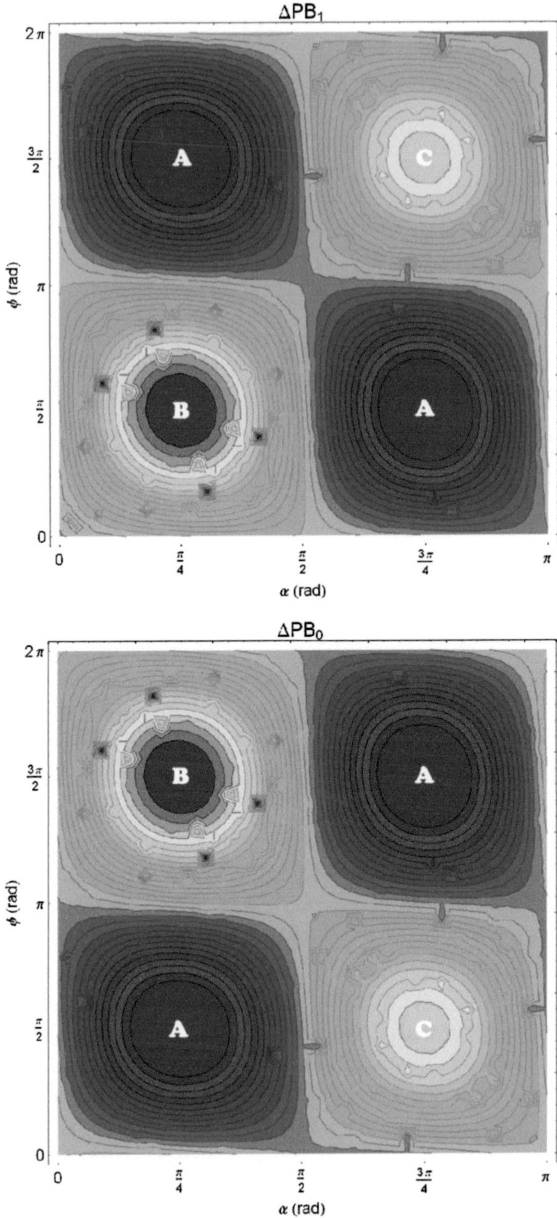


**Fig. D.10** Difference between numerical singles probabilities and evaluations of Eqs. D.37 and D.38. Here the regions labeled "**A**" reach minima of $5.7 \times 10^{-7}$, the regions labeled "**B**" reach maxima of $6.08 \times 10^{-6}$, and the regions labeled "**C**" reach maxima of $5.51 \times 10^{-6}$. Small blotches indicate regions in which numerical integration has produced larger errors

## References


1. W. Heisenberg, Reminiscences from 1926 and 1927, in reference [35]
2. P.J. Bussey, Phys. Lett. A **123**, 1–3 (1987)
3. K.A. Peacock, Phys. Rev. Lett. **69**, 2733 (1992)
4. J.B. Kennedy, Phil. Sci. **62**, 543–560 (1995)
5. P. Mittelstädt, Ann. Phys. **7**, 710–715 (1998)
6. K.A. Peacock, B.S. Hepburn, in *Proceedings of the Meeting of the Society of Exact Philosophy* (1999), arXiv:quant-ph/9906036
7. S. Weinstein, Synthese **148**, 381–399 (2006)
8. W. Heisenberg, The development of the interpretation of quantum theory, in Niels Bohr and the Development of Physics, ed. by W. Pauli (Pergamon, London 1955)
9. Th. Görnitz, C.F. von Weizsäcker, Copenhagen and transactional interpretations. Int. J. Theor. Phys. **27**, 237–250 (1988)
10. N. Bohr, *Atti del Congresso Internazionale dei Fisici Como*, 11–20 Settembre 1927, vol 2 (Zanchelli, Bologna, 1928), pp. 565-588
11. N. Bohr, Discussions with Einstein on epistemological problems in atomic physics, in *Albert Einstein: Philosopher-Scientist*, ed. by P. Schilpp (Open Court, 1949)
12. H. Everett, III, Rev. Mod. Phys. 29, 454 (1957); see also [35]
13. J.A. Wheeler, Rev. Mod. Phys. 29, 463 (1957); see also [33]
14. D. Deutsch, *The Fabric of Reality: The Science of Parallel Universes—And Its Implications* (Penguin Books, London, 1998). ISBN: 978-0140275414
15. Carver Mead, Collective Electrodynamics (The MIT Press, Cambridge, 2000); ISBN: 0-262-13378-4
16. R.E. Kastner, *The Transactional Interpretation of Quantum Mechanics: The Reality of Possibility* (Cambridge University Press, Cambridge, 2012)
17. T. Maudlin, *Quantum Nonlocality and Relativity*, (Blackwell, 1996, 1st ed.; 2002, 2nd. ed.)
18. J. Berkovitz, On causal loops in the quantum realm, in *Proceedings of the NATO Advanced Research Workshop on Modality, Probability, and Bell's Theorems*, ed. by T. Placek, J. Butterfield (Dordrecht, Kluwer, 2002), pp. 233–255
19. R.E. Kastner, Cramer's transactional interpretation and causal loop problems. Synthese **150**, 1–40 (2006)
20. L. Marchidon, Causal loops and collapse in the transactional interpretation of quantum mechanics. Phys. Essays **38**, 807–814 (2006)
21. J.G. Cramer, The transactional interpretation of quantum mechanics. Rev. Mod. Phys. **58**, 647–687 (1986)
22. J.G. Cramer, Found. Phys. Lett. **19**, 63–73 (2006)
23. E. Tammaro, Why current interpretations of quantum mechanics are deficient (2014, unpublished), arXiv:1408.2093v2 [quant-ph]
24. J.G. Cramer, G.A. Miller, J.M.S. Wu, J.-H. Yoon, Phys. Rev. Lett. **94**, 102302 (2005), arXiv:0411031 [nucl-th]
25. G.A. Miller, J.G. Cramer, Polishing the lens: I pionic final state interactions and HBT correlations—distorted wave emission function (DWEF) formalism and examples. J. Phys. **G34**, 703–740 (2007)
26. M. Luzum, J.G. Cramer, G.A. Miller, Understanding the optical potential in HBT interferometry. Phys. Rev. C **78**, 054905 (2008)
27. P.J. Lewis, Interpretations of Quantum Mechanics in the Internet Encyclopedia of Philosophy (2015), http://www.iep.utm.edu/qm-inter
28. R.E. Kastner, J.G. Cramer, Why everettians should appreciate the transactional interpretation, unpublished (2007), arXiv:1001.2867 [quant-ph]
29. S.S. Afshar, Violation of the principle of complementarity, and its implications, Proc. SPIE **5866**, 229–244 (2005), arXiv:0701027 [quant-ph]

30. A.C. Elitzur, S. Dolev, Multiple interaction-free measurements as a challenge to the transactional interpretation of quantum mechanics, in *AIP Conference Proceedings Frontiers of Time: Retrocausation—Experiment and Theory*, vol. 863, ed. by D. Sheehan (2006), pp. 27–44
31. N. Herbert, Am. J. Phys. **43**, 315 (1975)
32. M.A. Horne, A. Shimony, A. Zeilinger, in *Sixty Two Years of Uncertainty*, ed. by A.I. Miller (Plenum Press, 1990)
33. A.F. Abouraddy, M.B. Nasr, B.E.A. Saleh, A.V. Sergienko, M.C. Teich, Phys. Rev. A **63**, 063803 (2001)
34. H. Everett III, Rev. Mod. Phys. **29**, 454 (1957). See also [36]
35. A.P. French, P.J. Kennedy (eds.), Niels Bohr, A Centernary Volume (Harvard University Press, Cambridge, 1985)
36. J.A. Wheeler, Rev. Mod. Phys. **29**, 463 (1957). See also [34]

# Index


**A**
AAD predictions, 94
Absorber, 3, 59, 60, 64, 69, 94, 107, 110, 112, 121, 142, 164, 165, 178, 182
Actual experiment, 75
Advanced solution, 1
Advanced wave, 1, 4, 50, 61, 179
Advanced-wave echo, 4
Afshar experiment, 111
Afshar, Shariar S., 111
Aharonov, Y., 94
Albert, D.Z., 94
Alternative interpretations, 6
Angular frequency, 169
Angular momentum, 5, 12, 13, 29, 31, 32, 39, 40, 43, 49, 92, 125, 156, 170, 171
Angular momentum conservation, 5, 30, 41, 113, 138, 171
Angular momentum quantization, 15, 17, 170
Annihilation radiation, 29
Antimatter, 50, 162, 172, 186
Arrow of time, 3, 53, 161, 178
Aspect, Alain, 35
Atom excited state, 66
Atom ground state, 66
Atomic nucleus, 11
Atomic spectra, 9, 12–14
Atomic transitions, 18, 66
Avalanche, 67, 110, 182

**B**
BBO crystal, 113
Beat frequency, 66
Becquerel, Henri, 40
Bell inequalities, 5, 33, 91
Bell inequality violation, 91
Bell, John Stuart, 29, 30
Bell state, 32, 122, 127, 148
Bell's theorem, 5, 30, 32–35, 37, 39
Bertel, Annemarie, 17
Bester, Alfred, 152
Beta-decay, 3, 29, 40, 41
Biedenharn, Lawrence C., 1
Black hole information paradox, 130
Block universe, 111, 165, 166
Blocked interference, 138
Bohm, David, 30, 42
Bohr complementarity, 25, 59, 70, 111, 175, 176
Bohr complementarity principle, 177
Bohr Institute, 14, 17, 23
Bohr model, 12, 14, 17
Bohr, Niels, 12, 14, 19, 20, 22, 23, 25, 27, 28, 40, 41, 57, 67, 177
Boltzmann H-Theorem, 163
Boltzmann, Ludwig, 163
Bombelli, Rafael, 47
Born probability rule, 25, 65, 68, 70, 109, 176
Born, Max, 15, 25, 57
Bose-Einstein statistics, 50, 172, 186
Boson, 49, 64, 172, 186, 190, 194
Boundary condition, 1, 3, 52, 65, 69, 72, 79, 85, 90, 91, 94, 98, 110, 156, 172, 183, 199, 202
Boxed-atom experiment, 117, 199
Bra, 32, 84, 102
Budrys, Algis, 152

**C**
Cardano, Gerolamo, 47



J.G. Cramer, *The Quantum Handshake*, DOI 10.1007/978-3-319-24642-0

Carlsberg Brewery, 14
Causality, 1, 51, 52, 96, 137, 164
Charge conjugation, 162
Circular polarization, 30, 86, 87, 96, 171
Classical limit, 170
Clauser, John, 35, 91
Clockwise, 86
Collapse, 24, 26
Complementarity, 25, 111, 176, 177
Complementary, 70
Complementary variables, 20
Complex conjugate, 25, 32, 63, 69, 87, 96, 176, 181, 185
Complexity, 24, 49
Compton effect, 40
Compton, Arthur H., 40
Confirmation wave, 63
Conjugate variables, 25
Conservation laws, 39, 41, 43, 44, 71, 72, 125, 129, 130, 164–167, 173, 178, 180, 183, 184, 197
Constructive interference, 76
Contra-factual definiteness (CFD), 94
Copenhagen, 14
Costa de Beauregard, Olivier, 5
Counter-clockwise, 86

D
D'Amato, S., 94
Davies, Paul, 5
Davisson-Germer experiment, 14
De Broglie guide waves, 5, 117
De Broglie wavelength, 13, 14
De Broglie, Louis, 5, 13, 17, 83
Debye, Peter, 17
Decoherence, 157
Decoherence time, 157
Delayed-choice experiment, 88, 98, 99
Destructive interference, 68, 76, 77, 103, 119, 121, 143, 145, 146, 173, 202
Determinism, 111, 165, 166
Deutsch, David, 158, 177
Diffraction, 76, 114, 115, 154, 173
Dipole moment, 66, 69
Dipole resonators, 66, 67
Dirac bra, 32, 84, 102
Dirac equation, 49, 64, 171, 185, 194
Dirac ket, 32, 84, 102, 152
Dirac, Paul, 19, 49
Distinguishable, 34, 76, 100, 112, 143, 144, 147, 177, 187
Disturbance model, 22
Dolev, S., 118, 120, 121, 124
Don't-ask-don't-tell, 8, 70
Dopfer experiment, 114
Dopfer, Birgit, 114
Down-conversion, 98, 113, 114, 137, 141
Duality, 24–28, 41, 57, 88, 111, 172

E
EEmitter, 110
Eigenvalues, 17, 60, 185, 191, 192
Einstein, Albert, 10, 27, 28
Einstein bubble paradox, 4, 78, 173
Einstein clock paradox, 27, 177
Einstein-Podolsky-Rosen (EPR), 5, 28, 29, 174
Electron, 171, 172
Elitzur, Avshalom, 100, 117, 118, 120, 121, 124
Elitzur-Dolev 2-atom experiment, 120
Elitzur-Dolev 3-atom experiment, 118
Elliptical polarization, 125
Emitter, 3, 52, 59, 60, 62–64, 66, 68, 69, 82, 94, 121, 178
Encryption, 155, 157, 158
Energy exchange, 77
Energy operator, 192
Entangled photons, 98
Entanglement, 29, 43, 152, 167, 174
Entanglement-coherence complementarity, 137, 207
EPR experiment, 5, 91
Euler, Leonhard, 47
Euler's formula, 47
Everett III, Hugh, 177, 186
Expanding universe, 3
Expectation value, 194

F
FAQ, 169
Fast Fourier transform (FFT), 20
Fermi-Dirac statistics, 50, 171
Fermi, Enrico, 41
Fermion, 49, 64, 171, 186, 190, 194
Feynman, Richard, 2, 51, 75, 156
Firewall, 130
Fourier analysis, 20, 176
Fourier transform, 20
Free will, 165
Freedman, Stuart, 35, 91
Freedman-Clauser experiment, 26, 35, 39, 91, 122, 125, 144, 183, 198

**G**
Göttingen, 15
Garg, A., 125
Gaussian pulse, 20
Gedankenexperiment, 22, 27, 75, 80, 83, 89, 116, 118, 120, 121, 124, 173
Ghost interference experiment, 113
Gisin, Nicolas, 35, 128
Gluon, 172
Gribbin, John, 6

**H**
Hahn-Meitner Institute, 6
Half-wave plate, 76, 137, 141
Hanbury-Brown-Twiss effect, 92, 183
Handshake, 3–5, 61, 63, 100, 124, 126, 130, 152, 158, 165–167, 180, 181
Hardy, Lucien, 117
Hardy one-atom experiment, 117, 199
Harmonic oscillator, 18
HBT effect, 92
Heisenberg uncertainty principle, 70, 152, 176
Heisenberg's microscope, 22
Heisenberg, Werner, 7, 14, 20, 22, 24, 26, 29, 57, 78, 81, 85, 175
Helgoland, 16
Herbert, Nick, 34, 148, 183
Hidden variables, 42
Hidden variable theories, 4, 30, 35, 42, 187
Hierarchy, 67, 68, 108, 121, 182
Higgs boson, 172, 186
Hilbert space, 44, 71, 183
Hoyle, Fred, 5
Hydrogen atom, 13

**I**
I, 47
Identity, 24
Indeterminism, 24
Indiana University, 3
Institute for Advanced Studies, 28
Integer spin, 172
Intensity interferometry, 92
Interaction-free measurement, 100, 117, 118, 120
Interference, 173
Interference complementarity, 142, 144
Interference suppression, 78
Internet Encyclopedia of Philosophy, 186
Interpretation, 57, 66
Interpretational problems, 6, 19, 24, 168
Interpretation, Bohm-de Broglie, 42, 65, 66, 93, 117
Interpretation, collapse, 117
Interpretation, consistent histories, 117
Interpretation, Copenhagen, 6, 8, 24, 57–59, 69, 75, 100, 112, 131, 175
Interpretation, decoherence, 85, 167
Interpretation, Everett-Wheeler, 101, 112, 129, 158, 177, 186
Interpretation, Ghirardi-Rimini-Weber, 66
Interpretation, knowledge, 4, 26, 58, 70, 78, 79, 84, 109, 121, 131, 167, 176, 186
Interpretation, many-worlds, 101, 112, 129, 158, 177
Interpretation, possibilist transactional, 71, 167, 183
Interpretation, Transactional, 6, 8, 57, 59, 68, 69
IQOQI, 125

**J**
Jordan, Pascal, 16, 20

**K**
Kastner, Ruth, 6, 67, 71, 108, 124, 182, 183, 186
Kelvin, Lord, 9
Ket, 32, 84, 102, 152
Klein-Gordon equation, 4, 64, 172, 185, 194
Konopinski, Emil, 54
Kramers, Hendrik "Hans" A., 14, 40
Kwiat, Paul, 105

**L**
Lamb shift, 52
Left circular polarization, 86
Leggett, Anthony J., 125
Leggett-Garg inequalities, 125
Lewis, Gilbert N., 77
Lewis, Peter, 68, 108, 186
Light polarization, 30
Light waves, 10
$LiIO_3$ crystal, 114
Linear polarization, 30, 86, 171
Local hidden-variable theories, 33
Localization, 190
Localized waves, 20
Logical positivism, 8, 17, 24, 58
Loopholes, 43
Low-intensity two-slit experiment, 76

**M**
Mach, Ludwig, 101
Magnetic flux quanta, 157
Malus' law, 31, 34, 106, 125
Matrix diagonalization, 17
Matrix element, 17
Matrix inversion, 17
Matrix mechanics, 17, 18
Maudlin *gedankenexperiment*, 67, 108, 182
Maudlin, Tim, 67, 108, 182
Maxwell, James Clerk, 10
Maxwell's equations, 1
Mead calculation, 68
Mead, Carver, 66, 110, 179, 182, 186
Measurement, 24
Measurement problem, 24, 26, 70, 83, 186
Modern physics, 4
Momentum entanglement, 114
Momentum operator, 192
Muon, 171
MWI, 177
MZ interferometer, 101

**N**
Narlikar, J., 5
National Security Agency, 155
Neutrino, 41, 171
Newton's 2nd Law, 57
No self-interaction, 51
No-signal theorems, 135
Non-commuting polarizations, 86
Non-commuting variables, 85
Nonlocality, 5, 6, 24, 28, 29, 39, 42, 91, 152, 167, 174
Nonlocal realistic theories, 124
Nonlocal signal, 37, 114, 115, 135, 174
Norman, Eric, 4

**O**
Observables, 25
Observer-created reality, 27
Observers, 70
Occam's Razor, 158, 181
Offer-confirmation echo, 63
Offer wave, 63
Operator, 173, 191

**P**
Paradox, 6, 58
Parallel universes, 158, 177
Parity, 3, 29, 31, 32, 50, 162, 171, 186
Parity nonconservation, 3
Parity transformation, 171
Particle in a box, 14
Particle-like, 172
Particle scattering, 18
Path-entangled EPR, 139
Path entanglement, 114
Path label, 76
Path labeling, 78
Pauli exclusion principle, 50, 171
Pauli, Wolfgang, 14, 41
Peierls, Rudolf, 26
Penniston, Penny, 6
Pflegor-Mandel experiment, 94
Philosophical economy, 65
Photoelectric effect, 10
Photon, 10, 13, 19, 22, 27, 29, 30, 32, 40, 60, 67, 77, 78, 83, 86, 88, 91, 94, 96, 101, 170, 172
Photon cascade, 32, 91
Photon polarization, 37, 171
Photon scattering, 40
Pilot-wave model, 5
Pion, 94
Planck's constant, 10, 170
Planck, Max, 10
Podolsky, Boris, 28
Polarization basis, 30, 37
Polarization correlations, 91
Polarization-entangled EPR, 138
Polarization transformations, 87
Position-momentum complementarity, 22
Positivism, 8, 17, 24, 58, 70, 75, 131
Prime numbers, 158
Princeton University, 28
Pseudo-time, 64, 178

**Q**
QM formalism, 173
Quantization, 170
Quantum computer, 155
Quantum computing, 156
Quantum dots, 157
Quantum eraser experiment, 98
Quantum field theory, 54, 67, 135, 136, 179, 182
Quantum gates, 157
Quantum interpretations, 59, 174
Quantum jumps, 13, 67
Quantum liar paradox, 124
Quantum mechanics, 6, 169
Quantum mystery, 75, 77

Quantum poker, 197
Quantum reality, 126
Quantum system, 173
Quantum teleportation, 152
Quantum textbooks, 59
Quantum weirdness, 126, 131
Quantum wormhole, 130
Quantum Zeno effect, 105
Quark, 171
Quarter-wave plate, 100
Qubit read-out, 156
Qubits, 156

**R**
Radiative damping, 51
Radioactivity, 10, 40
Randomness, 24, 67, 110, 111, 139, 168, 182
Real 3D space, 71, 181
Realism, 126
Realism test, 125
Reality, 70
Relativistic wave equations, 64
Relativistic wave mechanics, 179
Relativity, 136
Renninger, 173
Renninger *gedenkenexperiment*, 83, 85, 108
Retarded potentials, 1
Retarded solution, 1
Retarded wave, 1, 4, 50, 60
Retrocausal, 37, 90
Reversibility, 164
Rice University, 1
Right circular polarization, 86
Rigid rotor, 18
Rosen, Nathan, 28
Rosenfeld, Leon, 27
Rotational symmetry, 360°, 50, 170
RSA encryption, 158
Rutherford, Ernest, 11

**S**
Sagnac source, 137
Schrödinger cat paradox, 80, 173
Schrödinger equation, 4, 17, 19, 64, 194
Schrödinger steering, 139, 142, 206
Schrödinger, Erwin, 7, 17, 19, 29, 67, 85
See-saw relation, 20, 137, 176, 190, 207
Self-energy problem, 50
Self renormalization, 54
Shaknov, I., 29
Shiekh, Anwar, 143
Shih, Y.H., 113
Shor algorithm, 157
Shor, Peter, 157
Signal + anti-signal, 142, 144
Slater, John C., 15
Snowden, Edward, 155
Solvay Conference, 27, 78, 177
Sommerfeld, Arnold, 17
Source of randomness, 110
Special relativity, 136
Spectral lines, 12
Spin, 49, 171
Spin $^1/_2$, 156, 170, 172
Spontaneous symmetry breaking, 67, 182
Spooky actions at a distance, 28, 37, 174
Square root of minus one, 47
Standing wave, 13, 64
Stark effect, 18
Stars, 9
Star Trek, 152
State vector, 69
Stern-Gerlach effect, 116
Stern-Gerlach separator, 116
Subluminal influences, 128
Superluminal influences, 128

**T**
Taylor, Sir Geoffrey Ingram, 76
Teleportation, 152
Testing interpretations, 175, 181
Thompson, J.J., 11
Time arrow, cosmological, 161
Time arrow, CP, 162
Time arrow, electromagnetic, 161
Time arrow, subjective, 161
Time arrow, thermodynamic, 161
Time reversal, 48, 162
Time-reversed EPR, 121
Time symmetry, 3, 50
Transaction, 63, 69
Transaction formation, 66, 67, 168, 182
Transaction model, 1-D, 59
Transaction model, 3-D, 62
Transactional handshake, 130
Two-slit experiment, 10, 75

**U**
Uncertainty principle, 20, 21, 43, 70, 109
Unitarity, 143
Universe wave function, 73
University of Washington, 3

**V**
Vaidmann, Lev, 100, 117
Van Vogt, A.E., 152
Variable entanglement, 137, 138, 141
V diagram, 5
Von Neumann, John, 84

**W**
$W^{\pm}$, 172
Wave, advanced, 1, 4, 50, 61, 179
Wave equation, 1, 17, 18, 173, 190, 193
Wave function, 4, 48, 173, 189, 190
Wave function collapse, 43, 78, 81, 83, 176
Wave-like, 172
Wave mechanics, 17, 18
Wave number, 169
Wave on string, 189
Wave, retarded, 1, 4, 50
Weak source, 77, 88, 102, 121
Wedge, 143
Wheeler's delayed-choice experiment, 88, 98, 99
Wheeler-Feynman absorber theory, 51
Wheeler-Feynman electrodynamics, 3, 51
Wheeler-Feynman handshake, 5, 54, 59, 68
Wheeler, John, 2, 29, 51, 177, 186
Which-way information, 76, 100, 111, 114, 115, 141
Wien, Max, 17
Wigner's Friend paradox, 82
Wigner, Eugene, 84
Wormhole connection, 130
Wu, C.S., 29

**Y**
Young, Thomas, 10, 75

**Z**
$Z^0$, 172
Zehnder, Ludwig, 101
Zeilinger, Anton, 98, 105, 114, 125, 137, 141
Zurich, University of, 17

Printed in the United States
By Bookmasters